The Role of Systems Methodology in Social Science Research

Dedicated to my father, Ruggiero,
and to the memory of my mother, Mary.

Frontiers in Systems Research:

Implications for the social sciences
Vol. 1

The objective of the series is to develop a rich resource of advanced literature devoted to the implications of systems research for the social sciences. The series includes monographs and collections of articles suitable for graduate students and researchers in academia and business, including rewritten Ph. D. dissertations. No undergraduate textbooks or reference books are included. Quality, originality and relevance with respect to the objectives of the series will be used as primary criteria for accepting submitted manuscripts.

The Role of Systems Methodology in Social Science Research

Roger E. Cavallo

State University of New York at Binghamton

Martinus Nijhoff Publishing
Boston/The Hague/London 1979

Distributors for North America:
Martinus Nijhoff Publishing
Kluwer Boston, Inc.
160 Old Derby Street
Hingham, Massachusetts 02043

Distributors outside North America:
Kluwer Academic Publishers Group
Distribution Centre
P.O. Box 322
3300 AH Dordrecht, The Netherlands

Library of Congress Cataloging in Publication Data

Cavallo, Roger E
The role of systems methodology in social sciences research.

(Frontiers in system research ; v. 1)
Bibliography: p.
Includes indexes.
1. Social science research. 2. Social sciences—Methodology. I. Title. II. Series.
H62.C3448 300'.1'8 78-10272
ISBN 0-89838-005-7

Printed in the United States of America.

Preface

While general systems research has had a considerable impact on research in the social sciences, this impact has been mainly conceptual and has not served to provide the operational and methodological aids for research which are possible. In addition, many of those systems-oriented directions and results which do impact social science research have developed independently and in piecemeal fashion in recent decades. The main development of this book is a cohesive framework within which to integrate results of general systems research and which provides a means for the organization of data and observations – and operational procedures by which to proceed – in the investigation and study of social and socio-technical systems.

The book systematically develops in the first five chapters all of the basic concepts and aspects which make up the framework, showing wherever possible the main sources of these concepts and placing them in historical perspective. The developments of the first five chapters are pulled together and integrated, in the last chapter, into a conceptual and operational general systems problem solving framework which extends the investigative capabilities of researchers of specific systems. The last chapter also contains an example of an overall investigation which utilizes the framework and which proceeds from system definition through the derivation of explanatory knowledge regarding the object system and which illustrates in detail most of the concepts and elements of the framework.

When the Society for General Systems Research was started, one of the founders – Kenneth Boulding – pointed to an inherent difficulty stemming from the fact that the same intellectual independence responsible for much of the vitality of general systems research could serve to prevent a consolidation and practical evaluation of many results and that this could in turn restrict the potential for further growth. He thus observed that one need which was most pressing, as well as most difficult to achieve, was the development of a viable overall framework which would be organically connected with all aspects of general systems research. Major progress toward achievement of this development has been the goal of the approach to general systems taken by George Klir. The process-oriented problem solving framework developed in this book is based primarily on the hierarchy of epistemological levels of systems which he has developed.

In this regard, I would like to express my appreciation for the motivation and inspiration supplied to me by George Klir, both directly and indirectly through his extensive and continuing work devoted to the creation and development of a perspective which encompasses, encourages and supports contributions from varied and diverse perspectives, and which provides means to integrate and benefit from this diversity.

I would also like to thank Walter Lowen for his administrative efforts devoted to the creation of an academic atmosphere, at the School of Advanced Technology, which is oriented toward and supports the integration of diverse approaches.

Lastly I would like to thank my wife Colleen for her support, especially regarding the tending of logistic details including the organization and typing of the manuscript for this book.

Contents

Introduction

In the introduction to his book *Causal Inference in Nonexperimental Research*, Blalock states that:

> We take the commonly accepted position that science contains two distinct languages or ways of defining concepts, which will be referred to simply as the theoretical and operational languages. There appears to be no purely logical way of bridging the gap between these languages. Concepts in the one language are associated with those in the other merely by convention or agreement among scientists.

The impact on Blalock of this position is to highlight a problem which the researcher in the social sciences must continuously face, and that is that he '*must, somehow or other, make sense out of his data.*'

The contention of this book is that, while Blalock's statements are basically correct, there are more facets to the problem than the single dichotomy theoretical/operational language implies and that the phrase 'merely by convention or agreement among scientists' unduly underemphasizes the role and importance of this convention or agreement. In this book we develop a methodological perspective and methodology which is based on general system considerations – that is, on the consideration of systems and systems properties as may derive from the study of phenomena in any field of investigation. This development leads naturally to a *general systems methodological framework* which is comprehensive and integrative, which takes advantage of the tension between theoretical and empirical concerns as a positive and dynamic aspect of the process of inquiry, and which in essence provides a means to unify these two concerns.

From one frame of reference it is possible to develop this framework as a metalanguage encompassing the theoretical and operational as used by Blalock. This would offer a pragmatist vantage point which 'bridges the gap' and provides a framework whose utilization offers the social scientist a way to 'make sense out of his data.' While this would constitute a worthwhile accomplishment, an alternate frame of reference will allow the achievement of this and more.

From this broader frame of reference, rather than focus on operational/theoretical language dichotomies, we use a somewhat different linguistic stratification which centers on degree of abstraction. This stratification is similar to one considered by Bunge (1969) and is based mainly on the notion of extensive or semantic referents.

In this regard we recognize that, in the investigation and development of general and systemic methodological principles, consideration must continually be given to the potential and the need for the determination of extensive referents by researchers with focused and developed expertise in the study of particular social and humanistic systems. From the point of view of utility to social scientists we recognize the need to organically connect this specific system expertise with methodological considerations in a manner which gives primary concern to the retention of the identity of specific systems. This concern is especially contradistinguished from the development of 'powerful' techniques whose use essentially demands the modification of the object of study.

The framework which we develop will, in addition to being useful in individual disciplines, provide an effective means for communication among investigators with expertise in different areas and, thus, provide an effective means for research related to objects, phenomena, and problems which transcend any particular discipline.

A major goal of the framework is thus to provide a device by which the social scientist may free his attention from an often oppressive concern with individual research techniques. Psychology – the study of the psyche; Sociology – the study of society; Anthropology – the study of man. Of course, no scientist in any of these disciplines can seriously believe, for example, that running a multiple linear regression of $n-1$ independent variables against a single dependent variable could constitute a study of the psyche, society, or man. Yet, because of the lack of an organized methodological approach through which a researcher could take advantage of general and comprehensive mathematical and abstracted operational procedures, the study of humanistic systems often degenerates into major concern with – and faddish attachment to – statistical procedures and/or contextually trivial formalisms.

The central development of this book is simple and is stated in terms of relations among:

1. social science disciplines (or disciplinary languages)
2. very general and abstracted disciplines (mathematics, computer science and philosophy)
3. general systems research.

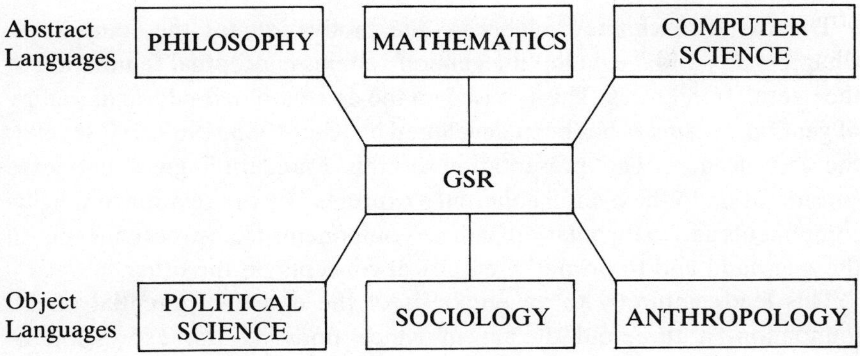

The main proposition is that general systems research provides an operational, conceptual, and logically consistent bridge between items listed under (1) and (2) above. Elaboration and development of this theme will evolve by considering various syntactic and semantic aspects of the general systems research component and its relation to the other parts of the diagram.

For these purposes we avoid possibly presumptive considerations associated with connotative aspects of General Systems *Theory*, as such, and concentrate on methodological aspects. The growing recognition and acceptance of general systems research (Bunge, 1977) has important ramifications on the activities of investigators in all of the more traditional, more semantically interpreted disciplines. The impacts are, thus, also significant for the clarification of the ontological and epistemological bases underlying investigative activity.

For preliminary consideration of the issues involved we refer to Kenneth Boulding's definition of General Systems as 'a level of theoretical model-building which lies somewhere between the highly generalized constructions of pure mathematics and the specific theories of specialized disciplines' (Boulding, 1956b). This simple but important characterization juxtaposes activities at two extremes: at one, these take place without regard as to whether objects of study have interpretation in the 'real world' – the abstracted products of dissociated reason in a sense become the reality; at the other, abstracted considerations are made only to the extent that the resultant theories afford interpretation in terms of an accepted reference class. In the development which follows we consider 'highly generalized constructions' to include metaphysics and computer science as well as mathematics and develop the methodological framework in this context.

The first two chapters elaborate the motivation for the framework. Chapters 3, 4 and 5 develop the general systems conceptual foundation of the overall framework. This is based on the epistemological level hierarchy of general systems as has been developed by Klir (1965, 1969, 1978c) over the last decade. The presentation of this foundation gives extensive consideration to the overall coherence expressed by the relation among its components and to the relation of these components to abstract concepts on the one hand and to primarily empirical concepts on the other.

This leads naturally to an emphasis of the dynamic potential of the foundation, a threefold dynamism which finds natural expression as movement and process among components as defined by the three relations.

The first part of Chapter 6 organizes and integrates the material and considerations of the previous chapters as a framework which embodies interactive aspects between an investigator and an environment and between an investigator and the framework. This interaction is expressed as an ongoing process of problem formulation and problem solution. The framework which is defined is thus termed *general systems problem solver.* The last part of Chapter 6 then demonstrates the applicability and utility of the framework through its use in an overall investigation related to the study of conflict within nations.

1. Science and Scientism

From the first quarter of the eighteenth century... as against the previous period, an emphatically one-sided rationalism appeared... Attention was directed much more to single facts.

Friedrich Klemm
A History of Western Technology

It was the life of the little day, the life of little people. And the man who had died said to himself: Unless we encompass it in the greater day, and set the little life in the circle of the greater life, all is disaster.

D. H. Lawrence
The Escaped Cock

1.1. The Social Sciences and the Analytical Method

To develop an appreciation for the role of general systems methodology in respect to the social sciences, it is necessary to trace some aspects of their development, and also to examine the status of certain methodological questions within them. By observing a page (or several fragments of several pages) of the history of science, we find that in the eighteenth and nineteenth centuries the forerunners of what we recognize today as the social and behavioral sciences, study areas such as psychology, economics and political science, were closely related with humanistic and metalinguistic branches of study. The natural sciences (actually, the sciences of the inert) were in the heyday of their period of accomplishment, exhibiting unparalleled capacity for 'dominion over the earth and mastery of nature.' A good portion of this accomplishment was undoubtedly attributable to a growing freedom from the hegemony and dogmatism of various forms of institutionalized thought. One early adherent of science, for example, Bernard de Fontenelle, in championing science's cause, extolled its principle values – skepticism, pragmatism, and relativism (Leiss, 1972, p. 77).

It is not surprising that, in parallel with the achievements of science, its principle approach, that of the analytical method, was to become quite thoroughly developed and articulated. Neither, unfortunately, is it surprising that this method would in turn become dogmatized and institutionalized. Succeeding centuries have found that the advances attributable to the analytical method have led many – sometimes implicitly, often quite explicitly – to promulgate the position that this method and associated mechanistic explanations were the *only* way to true knowledge. Though this position has since been called into question, if not abandoned, by most physicists (Heisenberg, 1958; Bohr, 1963), significant effects of its influence may still be found in the considerations of the social and behavioral sciences. Young (1976) quotes Oppenheimer's warning that, 'the worst of all possible misunderstandings would occur if psychology should be influenced to model itself after a physics which is not there any more, which has been quite outdated' and it is possible to conclude (Hayek, 1955, 1974) that advances in all the social sciences have been retarded by struggles stemming from aspirations to and emulations of the limited mechanistic ideal (or, as Hayek observes, from imitation of methods rather than the spirit of science).

It is worth elaborating some of the issues involved since they have a direct bearing on the evolution of general systems research as well as its status with respect to research in the social sciences. In the interest of clarity it serves to somewhat concentrate the discussion here on one branch of the social sciences, economics, and to relate these remarks to more general considerations in succeeding chapters.

1.2. Hucksterism in Economics

While the following characterization of great economists may not be the fairest, it has been made by one of the major historians of economics and it underscores a characteristic which is relevant to the development of the systems movement. Stigler characterizes the successful economist as one accomplished in 'hucksterism… a one-sided man… He is utterly persuaded of the significance and correctness of his ideas and he subordinates all other truths because they seem to him less important than the general acceptance of his truth' (Stigler, 1955). Stigler is referring in particular to Jevons, whose major work, originally published in 1871, was a significant contribution to the development of marginal utility theory, a development which in turn prepared the way for the mathematization of economics. Jevons himself stated that, 'Economics, if it is to be a science at all, must be a mathematical science.' He was clearly attempting an imitation of Newtonian mechanics ('The theory here given may be described as the mechanics of utility and self-interest'), and was one of the first to treat factors such as population, theretofore considered as inseparable from economic considerations, as exogeneous variables. It is apparent that in the days of so-called 'literary' economists such innovations played an important role in emphasizing the potentials for abstraction in the social sciences. It also appears, however, that this forward movement induced expectations for greater achievement than could be realized and that oversubscription to the proposed methods has essentially resulted in the de-socialization of economics as a social science.

The enchantment of economists with the methods of natural science stemmed at least from the middle 1700's with the possibilities referred to by Condorcet, a mathematician turned political theorist and activist:

> The only foundation for the natural sciences is the idea that the general laws, known or unknown, which regulate the phenomena of the Universe, are necessary and constant; and why should that principle be less true for the intellectual and moral faculties of man than for the other actions of nature (In Hayek, 1955).

Illustrative of the impact of such thought is the fact that the quote was used as the motto for one of the books of J. S. Mill, who also wrote *Principles of Political Economy*, the 'undisputed bible of economics all through the second half of the nineteenth century' (Blaug, 1962, p. 163).

In a significant study, *Divided Existence and Complex Society*, van den Berg explores and discusses from a historical perspective many of the issues and themes relevant to the analytic division of the perceived world (and to the accompanying belief in the possibility of a strict separation between subject and object). Of interest here is his dating the desire – the need – for separation from the 'marvellous, amazing, astounding *all*', from 'the totality' – in order to be able to conduct scientific investigation – to 1740 and to Abraham Trembley's removal of a polyp from its environment in order to be able to observe it free from the wonder of its natural habitat. He quotes Trembley, who gives a description of the necessity to restrain astonishment and amazement, to take an object of study – 'which belongs to the totality so emphatically' – and 'remove it from this totality' in order to be able to observe and, further, to *count* that which one observes. Van den Berg's compelling considerations of these issues in the eighteenth and nineteenth centuries may be considered somewhat overdone, to be able to argue for as precise a temporal identification of ideas as he does. Nevertheless, it is impossible to not recognize the importance of the period – the period in which 'scientific' economics also had its start – for the growth of the analytic method.

We will refer below – in consideration of the development of the cybernetic paradigm – to a more general impact during this period of analytic and reductionist tendencies, but here we further indicate their effect on economics through consideration of the parallels in the following statements:

> We cannot conceive any further explanations to be either necessary, desirable or possible for as soon as we know what is meant by the words configuration, motion, mass and force, we see that the ideas which they represent are so elementary that they cannot be explained by anything else (James Clerk Maxwell, quoted in Makridakis [1977], who observes that, 'A few years later, of course, his words crumbled').

> Its [economics] ultimate laws are known to us immediately by intuition . . . its method is as sure and demonstrative as that of kinematics and statics (Jevons, in Stigler [1965]).

> The postulates from which economics developes its propositions are so much the stuff of our everyday experience that they only have to be stated to be recognized as obvious (Robbins, 1935).

For an adequate overall perspective it is important to be aware, however, that even within economics inherent contradictions and potential difficulties associated with extreme affiliation to analysis based methods did not go unrecognized. This is evidenced, for example, by the Austrian economist Karl Menger, whose book *Grundsätze* was also published in 1871 and independently introduced the marginalist approach. Menger was reticent to accept mathematical formulations as he felt they were of no use in helping economists to get at the qualitative 'essence' of the phenomena they were supposed to be studying. The variance between the two positions would lead to more than one period of *Methodenstreit*, but it is generally agreed that that of Jevons, the search for 'immutable laws of nature' – or, as characterized by Hayek (1955), the 'scientistic hubris' resulting from the 'cult of Newton' – would be granted more general primacy among economists.

This is not to say that the sentiments expressed by Menger's position have ever been totally expunged from within economics. Evidences of a lingering dissatisfaction can be found in the writings, for example, of Veblen (1919), Leontief (1966), Rotwein (1962), Samuelson (1963), Boulding (1950), Papandreou (1958), Georgescu-Roegen (1966), and Schumacher (1973), and can be detected in the following quotations: 'If a theory is the best simple theory in town, that is no excuse for saying it is a good theory if it is not a good theory' (Samuelson, 1963); 'I agree, too, that by itself economics affords no solutions to any of the important problems of life' (Robbins, 1935); 'The true output of this economic process is not a physical outflow, but the enjoyment of life. Without recognizing this fact and without introducing the concept of enjoyment of life we are not in the economic world' (Georgescu-Roegen, 1966). It is also interesting to consider the comment from the introduction of *An Introduction to Positive Economics*: 'Philosopher friends have persuaded me that, when pushed to its limits, the distinction between positive and normative becomes blurred, or else breaks down completely' (Lipsey, 1963). These works and sentiments have had little effect, however, on what is recognized as hard-core or 'scientific' economics.

Taking a more charitable perspective, it is not difficult to understand that the lack of alternative conceptual and methodological apparatus – as well as the greater simplicity in incorporating methods rather than scientific spirit – could lead to attempts to apply methods to problems for which they were not intended; and it is not difficult to understand the result – the conceptual modification of situations under investigation to fit tools and models available.

It is necessary to emphasize here that the value of this differentiation of scientific consciousness which these abstractions represent is not being questioned – nor even its necessity; that the original emphases on analyticity and the introduction, into the study of social questions, of caeteris paribus as an analogue to the laboratory methods so vividly described by Trembley are not of themselves alarming, but rather the excesses which developed from extreme enchantment with this methodological orientation. Jevons, for example, at the time of the first edition of his book, though treating population as an exogeneous variable, fully ascribed to the population theories of the time and at first expressed the need for a balanced view regarding the possibility or validity of achieving numerical expressions for his main ideas. These reservations were removed from later editions, however. Evidently, the enthusiasm of the times for the division and domination of nature, the simplicity of hucksterism, the scientistic hubris were too appealing and abstracted and separated parts became so many little realities.

1.3. The Systems Paradigm and the Re-emphasis of Interaction

It is important for those domains of social science which still retain an interest in both the social and the science of social science to recognize the resultant object-modification. The point is especially relevant since there are beginning to appear – in the literature of political science and sociology, for example, study areas which have to a certain degree resisted the lure of the apparent successes of an uncritical disregard for interaction and the whole – comments referring to the 'coming of age' of economics as a science, and to the 'towering edifices' which certain economic theories supposedly represent, comments very similar to those made in the eighteenth and nineteenth centuries by economists in simplistic emulation of the natural sciences.

It is the contention of this book that a necessary step in avoiding the extension of this process is the development of the means whereby the achievements of the scientific process – of conscious rational activity represented by *critical* abstraction, categorization, and determination of general principles – can be utilized while at the same time retaining the identity of the object of study.

A topic in the sociology of the social sciences – or in the study of human folly – which seems worthy of the deepest investigation is: how or why the caricature of classical science, which has supplied a basic motivation for the systems movement, is so relevant. Alternatively, why have the emulative social sciences not attempted emulation of any of the developments within physics since at least the special theory of relativity. This observation leads to an alternative formulation of the theme of this thesis and of the rationale for the study of systems, that being, how can insights of modern science – say physics, which are so intricately connected to a jargon and reference frame which has developed over hundreds of years, and which would be extremely difficult to translate directly to alternate linguistic reference frames – how can such insights be made to play a meaningful role in social science research? Further, is it possible to develop a mechanism by which all sciences can keep up with further advances in the overall body of knowledge which scientific inquiry should represent?

The most significant expression of the uneasiness with the exaggerated reliance on the analytical method was the recognition by various scientists in the 1930's and 1940's of this century that the idea of 'system,' which was at least as old as that of the traditional disciplines, and which represented the antithesis of the ability to study parts in isolation, was a concept worthy of study in its own right.

The argument by which the study of *systems* developed its momentum was fairly simple:

- Study by division and separation, by specialization and disciplinary categorization has evolved to a far greater degree than seems justified.
- Many of the ills which it seems that science should not have allowed to happen have happened; many of the answers which it seems that science should have been able to give have not been given ('... it has contributed scarcely anything to our understanding of social phenomena' [Hayek, 1955, p. 14]).
- An implicit assumption – possibly true for certain of the simple systems which gave rise to the methods used – is that the results of the study of fragments of a fragmented whole could easily be combined together to constitute knowledge of the whole.
- It seems that this assumption is false. At the least it is not valid for most of the complex phenomena constituting the essence of humanistic, as opposed to mechanistic, objects of investigation (Zadeh, 1974). Major characteristics of such phenomena have precisely to do with the ways in which fragments are related, rather than how they behave in isolation, or even what they are in isolation.
- The notion of system explicitly entails emphasis on potential interactions among all the parts that are perceived or distinguished.
- Disciplinary study must be augmented by a much more sophisticated knowledge of the effects which accrue specifically to the existence of interactions among parts of a recognized whole, that is, by the study of systems.

The argument was in fact so simple, straightforward and compelling, and did in fact seem to so crystallize an idea as old as the history of inquiry, that it generated a great deal of enthusiasm and support from investigators in many disciplines, particularly those of the biological and social sciences. A great deal of research has been generated which is devoted to an elaboration of the basic idea within the context of different disciplines (e.g., Buckley, 1967, 1968; Berrien, 1968; Pattee, 1973; Boulding, 1955) as well as to the explication of the various reasons that the systems concept has been accepted to the extent that it has (e.g., von Bertalanffy, 1968) and to the elaboration of *system* as an acceptable field of study in its own right (Klir, 1969).

1.4. The Status of Mathematical Research

An important factor which has also contributed to the motivation of systems research, which is rarely mentioned, and which is probably the most important in that its consequences are easily specifiable, is the extreme development of mathematics as a study independent of sense observations or any 'other reality than symbol manipulation' itself. The nineteenth century, 'the most revolutionary in the history of mathematics,' witnessed the rise of abstract algebra represented by De Morgan's insistence that 'with one exception, no word or sign of arithmetic or algebra has one atom of meaning throughout this chapter, the object of which is symbols and their laws of manipulation' (in Boyer, 1968). This view was quite different from that held to varying degrees until this time – that mathematics was a formal expression of observations of nature and sense experience as, for example, calculus was developed in the context of fluidity and motion. While operating in such a framework it is reasonable to presume that mathematical developments would have been more integrated with, attuned to, and suited for observations whose description was desired, including those from the social sciences.

It is true that the enthusiasm represented by formalist views is not what it was at the turn of the twentieth century, but the autonomy of mathematical studies has persisted and mathematics has taken great advantage of this freedom. There thus resulted considerable new mathematical developments, developments not tied to 'real-world' referents. A positive motivation for general systems research resulting from this process stems from the fact that these developments constituted a significant set of representational forms whose semantic extensions had not yet been developed, and which were available for use by scientists devoted to a redress of the lack of study of interaction.

Within the framework of the original enthusiasm for the study of systems, the expectation of achievement through utilization of these mathematical developments was naturally very high. Subsequent experience suggests that the extent of the difficulties was underestimated regarding both (a) the achievement of non-trivial results of direct utility having to do with this new focus, and (b) the (closely related) generation of useful results which possess both the power and clarity of abstraction and the meaning and significance which can only be associated with objects of some direct concern. Of course, (b) is merely a restatement of a fundamental problem which was recognized in the consideration of economics and which is the dilemma which must continue to be faced in all juxtapositions of one and many.

A resolution of this dilemma is suggested, however, through the elaboration of secondary but more profound effects of systems research, a resolution which constitutes a reorientation of our view toward this dilemma. Before dealing with this, though, we must mention the negative motivation for systems research which is a function of the autonomy of mathematics.

As mathematics developed less and less concern for 'single atoms of meaning,' object-oriented studies were left to a large degree to make do with what they had (which in the case of the social scientists came to them once removed from usages in mechanistic systems) or to utilize a quite different methodological orientation than that of developing mathematical abstractions directly from phenomena. We may loosely refer to this orientation as an applied mathematics methodology. Under this mode of operation the scientist is left to his ingenuity, basically to adapt his object of study to fit the existent mathematical constructs, or from the other perspective, is left to the mercy of the mathematician disenchanted with the purity of mathematics, and who is unable to accept the heritage of the, at least honest, impotence implied by the nineteenth century 'revolution,' and who undertakes a search for an area of application. For certain well defined situations this produces satisfactory results, but there remain a substantial body of cases where it does not.

The situation has been clearly described by Kemeny: 'In this case the scientist must create a new branch of mathematics or must appeal to the mathematician to undertake this task for him.' Experience has, however, shown that the basic facts of the situation involving the responsibilities for the more difficult problems that the social scientist has agreed to face preclude meaningful matching of trivially simple abstractions to complex situations unless this involves a deep awareness of these situations that can only arise from long exposure and study of them. Kemeny continues: 'Many mathematicians have the impression that mathematical problems in the social sciences are entirely trivial. On the contrary, most problems in the social sciences are too difficult for present-day mathematics' (Kemeny, 1961).

There are occasionally situations, of course, represented by such developments as dynamic programming, input-output economics, linear programming, and the fast fourier transform, which take advantage of a peculiar knowledge and ability, on the part of certain personalities, to operate close to both ends of the object-abstraction spectrum referred to in the introduction. These cases are relatively rare however. Unfortunately, they invariably also represent the difficulty (for example, as with linear programming) that they in turn become subject to use as abstractions for situations different from those for which they were developed. This once again leads to modification of the objects of original concern – and of the problems as originally conceived, and which are not presently solvable, into those for which solution techniques exist.

Extended consideration of general systems research serves also as a viable alternative to this difficulty. The example of this history, though, clearly indicates the difficulty and folly of attempting the development of a general systems methodology intended to blanket 'real-world' phenomena and solve significant overall problems merely through the use of the methodology. Rather it points to the pressing need for a methodological framework which is derived from particular systems in general and varied contexts, which is based on those properties of systems which are indeed found (or are generally recognized as existing) in these varied contexts, and which *retains the ability for adaptation of results to fit specific contexts.* Consideration of this need is a major current motivation for general systems research.

1.5. General Systems Methodology and Empirical Research

While the emphasis on interaction of parts is a more specifiable motivating factor in the original generation of systems research, there were a number of others, all to some degree expressions of a dissatisfaction with dogmatic and bureaucratic aspects which had come to be recognized in parts of science. One of these recognized that the incredible proliferation of specialized areas of study, each one having associated with it its acceptable methodologies, had to involve – besides the likelihood of not being able to create a meaningful integration – collective inefficiencies in the sense that many insights, and even techniques, generated within an area could be useful in others. There was basically no mechanism to accomplish the cross-fertilization to take advantage of such situations, and whatever did occur would have to depend on chance discoveries by different investigators. We note again, however, that without a *general* motivation for the original discovery, then the better suited a conception or technique is to a certain referential situation, then the more integrally it is associated with that situation, and the more likely it is that the technique in question – when removed from the context or transferred to another context (Klir, 1972b) – will retain little of its pragmatic significance.

Kenneth Boulding – who has long been actively aware of both the advantages and limits of scientific activity and, in particular, those referred to in the foregoing remarks (Boulding, 1950, 1956a) – referred in 1953, in correspondence with Ludwig von Bertalanffy, to the possibility of a *general empirical theory*, the development of which could prove fruitful to investigators in many different disciplines. The implication of the observation is that – for any foreseeable conception of the nature of scientific activity – observation, data, and data processing must continue to play a significant role in that activity. Unfortunately, the emphasis on these aspects which has developed in each of many disciplines has unduly diverted attention from the proper foci of attention in those disciplines. Boulding suggested that the articulation of the general issues involved with respect to these aspects would serve to allow a refocus of disciplinary attention to issues relevant to the classes of phenomena which represent the disciplines' proper domains of investigation.

The general agreement on this and other important issues eventually culminated in the establishment of the Society for General Systems Research, which included representation by like-minded scientists in many of the disciplines. The major inspiration in the movement which the Society represented was supplied by Ludwig von Bertalanffy, a theoretical biologist. Though von Bertalanffy's initial vision was probably more 'theoretically' oriented it is clear that his perception, as well as Boulding's and others of the Society, also attached importance to the methodological emphasis (see von Bertalanffy, 1968).

A closely related issue which has also played a significant role in the motivation of general systems research has been implicit dissatisfaction with the position (and responsibility) of man that had been inherited from the traditional scientific dogmatic or metaphysical realism (cf. Heisenberg, 1958). For, if the 'truths' existed in nature and it was merely scientific man's objective to discover them, then either man is denegrated by the denial of a creative nature of his activity or, as reaction to this possibility, he must set himself up to achieve the mastery and domination of nature. In either case the resultant non-responsibility in the light of increasing potential for catastrophe is unacceptable. The creative, and responsible, aspects of scientific inquiry must be re-emphasized (cf. Leiss, 1972).

At this point, and in light of the purpose of this book as described in the introduction, we can interpret the foregoing remarks as an expression of the need which had developed for a re-integration of the basic components of scientific activity into a more adequate realization of a 'total' collective scientific self.

While such expressions have continually been made from a perspective independent of science, the general systems expression was especially significant because of its recognition of and continued relation to questions pertinent to scientific foci. This continued link to scientific questions was a necessary condition for advancement for, as Neuman has argued, 'romantic attempts to revalue or reverse this development [the degeneration of differentiation and specialization into overspecialization] necessarily result in regressions, because they take no account of its forward tendency and misunderstand its connection with the historically positive evolution of the ego and consciousness' (Neumann, 1954).

1.6. The Influence of Cybernetics

Concurrent with the formulation of these issues from a biological and social science orientation was a parallel movement by scientists working through insights primarily furnished from the area of cybernetics. Though current conceptions more closely identify cybernetics and systems science (Klir, 1970a), the original impetus for cybernetics was largely supplied by generalizations generated from technological processes, in particular the study of machines.

While there have been many analogies using the basic workings of machine which have been superficial, this does not diminish the significance which many of the associated concepts have had on the workings of science and, indeed, in our culture. The development and use of machinery and technology are intricately and deeply related to all aspects of our life and history and it is important that certain general facts be pointed out.

Tools have been in use by man since the beginning of his recorded history. The distinction between tool and machine is not absolute and clear-cut, but a useful differentiation that can be made is that the operator of a machine, as opposed to the user of a tool, does not take part in the actual work that the machine does. He is essentially separated from the process the machine is engaged in.

We have seen that the intellectual tenor of the eighteenth century – the century of rationalism – has had significant effect on the development of economics. It is generally agreed that this century also marked the transition – particularly with Hargreaves' spinning jenny in 1764 – from manual work to machine industry, the separation of man from the generation of his production (Mantoux, 1928; van den Berg, 1974; Klemm, 1954).

Equally relevant is that this mechanization put only the finishing touches on the centralization of industry which gained momentum at the turn of the century. Van den Berg describes the situation:

> Well then, we are trying to find an explanation for the obscure process of division which made its appearance in the eighteenth century... That this was the natural, habitual state of affairs does not seem likely... What was the origin of the fissiparous tendencies disrupting human existence? ... There is no earlier record of these phenomena because, again, they were not of all times, but made their appearance at this very moment (van den Berg, 1974).

The results of this centralization of industry were not only to separate man from the processes of production, but to consciously divide the production process into smaller parts which would be completed and assembled at a later stage. Thus the approach and feeling represented so forcefully by the analytical method was spread and distributed, in a fundamental way, to all aspects of society.

This infiltration into the source of livelihood was only to contribute to the weight of the dogmatic yoke of our intellectual history (the phrasing is from Bruner, et al., 1956), which included – along with naive realism and the subject/object separation – an extreme deterministic bias, one opposed to any 'break in strict causal sequence and succession' (Mowrer, 1960).[1] From

1. 'It stands to reason that the empiricist thinks causally, the necessary connection between cause and effect being take as axiomatic. The empiricist is oriented by the empathized object; he is, as it were, "actuated" by the external fact and impressed with a sense of the necessity of effect following cause. It is psychologically quite natural that the impression of the inevitability of the causal connection should force itself on such an attitude. The identification of the inner psychic processes with external facts is implied from the start, because in the act of empathy a considerable sum of the subject's activity, of his own life, is unconsciously invested in the object... So whenever the object predominates, an assimilation to the object takes place... The psyche then labors under the impression of the exclusive validity of the causal principle, and the whole armoury of the theory of knowledge is needed to combat the overmastering power of this impression... The determinism of the empiricist, therefore, is a foregone conclusion, provided that he carries his thinking that far and does not prefer, as often happens, to live in two compartments – one for science, and the other for the religion he has taken over from his parents or from his surroundings' (Jung, 1921, pp. 316-317).

It is interesting to contrast Mowrer's admission, forty years later, in a chapter entitled 'Imagery, Memory, and Attention', that: 'For psychologists whose training dates back to the 1920's or 1930's the terms which constitute the title of this chapter were then and perhaps are

a pragmatic perspective the resultant emphasis on sequential characterizations of process description had an unduly inhibitory effect on the investigation of the synchronic and mutually reinforcing aspects of processes. But this missing development is precisely where we return to the consideration of cybernetics. In the intervening two centuries from 1764, during which machines had become a fundamental and integral aspect of our lives, their study and utilization had reached a quite different degree of sophistication. Certain concepts, which were quite advanced compared to the naive mechanistic conceptions of the eighteenth century, had become fundamental and commonplace. One of these was *feedback*, which simply says that output of a process returns and is utilized within the process as input.

The seemingly innocuous technological utilization of feedback – and the related ideas of servomechanism and regulation – were quite well understood, but the important fact is that to the progress-oriented society which had developed there would be no question of the acceptability of the utilization of ideas which had a positive effect on our technological capability. The fundamental cybernetic characterization which arose from such ideas – and there are two aspects of consideration for which this is relevant, those dealing with the construction of machines and those dealing with the study and description of already constructed machines – represented the more interesting machines mainly in terms of a backward and forward flow of information, with the implication that for the parts of such machines, 'each is a cause, and each an effect, of the other' (Tustin, 1952). Though the history of thought has of course always been represented by attempts at explication of these ideas (e.g., 'The thinker makes a great mistake when he asks after cause and effect. They both together make up the indivisible phenomenon' [Goethe]), it was to a large extent the influence of machines – one of the contributing factors to the original break from primary concern with interest in the whole, to the growth of analytic consciousness – which was to also have a significant effect on the growth of reintegrative studies. 'Now, however, it has proved

---

still, in some measure taboo. Many of us were taught, under pain of banishment from professional psychology, never to use these terms, at least not during 'working hours.' In our discourse with laymen it was permissible... but such language was deemed completely unsuited to the purposes of science' (Mowrer, 1960, p. 163).

'Has it ever... been observed that a scientific truth needed to be elevated to the rank of dogma? Truth can stand on its own feet, only shaky opinions require the support of dogmatization. Fanaticism is ever the brother of doubt' (Jung, 1934).

possible to construct *machines* with impressive capacity for *self*-regulation; and since there can be no imputation of "subjectivism" here, such machines and the principles they embody have inspired new attempts to account for the adaptive capacities of living organisms' (Mowrer, 1960, p. 264).

One of the major cybernetic contributions to modern science has thus been an acceptable escape from a simple and narrow view of causality and determinism. The potential for development of the idea inspired a great degree of experimentation in almost all areas of inquiry and it would be difficult to exaggerate the conceptual impact of utilization of this freedom. While vestiges of the dogmatism still remain, much of the pedantic nature of what Eddington succinctly refers to as the 'old causal method' (Eddington, 1934, p. 73) has disappeared from most areas. It is interesting to contrast Eddington's account of the parallel situation in physics, which has always enjoyed special immunity to the rantings of fanatical objectivism. He likens the development of the science of physics to the utilization of paper currency under the gold standard – the causal law being the gold stored in the vaults. The progression in physics, although still thought to be dependent on the gold, eventually saw fewer and fewer cases where the gold was actually produced and this led to skepticism as to whether the gold in the vaults actually existed.

As Eddington points out, his analogy would carry much more drama if a check of the vaults showed them to be empty. The situation he describes, however, fits more meaningfully into the pragmatic framework we are trying to develop. The climax of Eddington's story is that the key to the vaults has been lost and it's uncertain if there's any gold or not. He ends the parable: 'But I think it is clear that, with either termination, present-day physics is off the gold standard' (p. 81).

The important point – the loss of the key (and many would argue that the key has been proven not to exist) – is not that resolution of the ultimate issues involved has been made, but that the very issues themselves have been determined to be the wrong, or at least not the most relevant, issues with respect to many of our important problems.

Even at its own level, however, there have been interesting approaches and attempts at resolution or clarification of the problem from a modern perspective (e.g., Brown, 1969; Varela, 1975, 1976). These are approaches which recognize the fundamental difficulty referred to by Cioran as the 'fall into time,' the trading of 'eternity for becoming.' The particular forms of the approaches are only tangentially relevant to the major objective here, which is pragmatically oriented and primarily methodological. It is necessary, however, to be aware of the potential significance of such attempts, which are predicated on subtle perceptions of wholeness, circularity, and self-reference and which, it seems, are basically experiential. The experience, however, is (by admission) universal and time-less and will thus continue to elude a description which is acceptable outside of the experience through means which are necessarily time-bound. Within context of the experience, however (whence no description should be necessary), it is – paradoxically – possible to enhance the feeling one has for the acceptability. Further, every such enhancement seems to intensify the experience. Conversely, it also appears to be possible to enhance the potential for attaining the experience through what, from the perspective of one-sided wholeness, might be considered impossible directions.

This point relates to complementarity and alternate modes of description or frames of reference (cf. Bohr, 1963) and implies with respect to complementary modes that any resolution must come from the postulation or search for metalinguistic expressions which would constitute a reference frame from which seeming contradictions may be embraced as part of a more encompassing perspective.

The situation has been stated from the perspective of physics by the Nobel physicist Wolfgang Pauli who offers that, 'It rests with the free choice of the experimenter (or observer) to decide... which insights he will gain and which he will lose... It does *not* rest with him, however, to gain only insights and not lose any' (Pauli, 1946; see also Pauli, 1952). Pauli offered these comments referring to physics, adding that the 'newer psychology,' referring to the framework of C. G. Jung, represented a potential escape from the inherent subjectivism and limitations of physics.

Pauli notes that just as the object of physics must reflect the observer, so too the processes of the unconscious 'cannot always be unambiguously ascribed to a particular subject.' Pauli offers his belief that:

> To anyone for whom a narrow rationalism has lost its persuasiveness, and to whom the charm of a mystical attitude, experiencing the outer world in its oppressive multiplicity as illusory, is also not powerful enough, nothing else remains but to expose oneself in one way or another to these intensified oppositions and their conflicts. Precisely by doing so, the inquirer can... approach to the poles of the antitheses... But contrary to the strict division of the activity of the human spirit into separate departments... I consider the ambition of overcoming opposites, including a synthesis embracing both rational understanding and the mystical experience of unity, to be the mythos, spoken or unspoken, of our present day and age (In Heisenberg, 1959).

Although considerations – speculations – such as these are often reserved for the last paragraph of works in which they are not the fundamental considerations, the importance of the issues involved seems to warrant them more than epilogue status. To balance the pomposity, however, it is worth considering Cioran's somewhat more cynical observation:

> A civilization begins by myth and ends in doubt... No civilization can begin by questioning values it has not yet created; once produced, it wearies of them and weans itself away... for the various beliefs it had engendered and which now break adrift, it substitutes a system of uncertainties, it organizes its metaphysical shipwreck (Cioran, 1964).

In contrast to the pessimistic disposition of Cioran's statement, the value-questioning which had become associated with cybernetic inquiry has constituted significant forward movement in many areas – for example in the study of evolutionary and cognitive processes. The latter are in turn especially relevant due to the insight which they provide for the study and characterization of collective cognition. Some of the most useful integration – of results from many other areas as well as those from cybernetic orientations – is represented in Piaget's studies on the development of intelligence, and of his extension, in turn, of these insights to the general process of inquiry. This process itself is basically cybernetic in character, but more specifically relevant is his characterization of the development of intelligence, the growth of knowledge, as a result of the interaction and mutual contribution of subject and environment. Certain of Piaget's results are important for the general theme of this book and will be referred to in that context.

Piaget's work, as is much of that which must be considered successful in other areas, has – and this is similar to the case of general systems – utilized cybernetic paradigms, but utilized them indirectly. There have also been direct attempts to achieve results through cybernetic approaches. We find these, for example, in the attempted development of machines which can learn. By attempting to create such machines we develop useful models which in turn contribute to the understanding of how we learn. While much has been accomplished, many of the original expectations were never approached and are now recognized as illusory. Compare Bar-Hillel's statement of 1952: 'If a human being can do it, a suitably programmed computer can do it too' with a later recognition that, 'It became more and more clear to me that man's ability to translate... rests on some kind of innate organization. At the moment there are no indications that we shall understand this innate organization in the foreseeable future' (in Fuchs, 1968).

The very recognition of this difficulty however, the refusal of hubris, coupled with the refusal to succumb to the ensuing doubt, offers the perspective from which it becomes possible to achieve both positive results and the reintegration of these results into progression toward a more comprehensive goal. While original expectations with respect to hard-core machinery have not been realized, the cybernetic characterization of system has effected a widespread and foundational shift in emphasis in the types of questions asked – and thus in the types of problems available for solution – by science.

Along with this shift, a more subtle, but equally important, contribution of the cybernetic orientation concerns a change in perception regarding the bifurcated status of the man/nature relation which has grown and intensified since the eighteenth century. This is directly related to the more sophisticated awareness and recognition of the role of technology in science and society. While in its original effects machine acted more as a barrier between man and environment, there has been a shift in emphasis toward the character of man as builder of the machine and hence as playing a major role in the structuring of the universe. The perception of technology has become – more properly – that of the study of technique, of technique of construction, of the ability to build a 'machine' to replicate perceived or desired phenomena.

Perhaps retention of the word 'machine' is unfortunate. We are no longer referring to pieces of hardware but to constructions of both abstract and concrete specification. The important connotation which 'machine' retains, however, is the implied inability to separate the construction from the source of input, an implication which removes from the domain of interest machines which are not to have 'one atom of meaning,' machines which don't function.

The extreme classical position with respect to the 'objectivity' of scientific knowledge and to the detachment of the observer had created what appeared to be insurmountable difficulties in developing 'scientific' social science. Ironically, the sophisticated emphasis and study of machines and machine constructability has significantly contributed to the awareness necessary to free scientists from the religious quest for (immutable) observer-independent laws. This freedom represents both the elevation of man, through recognition of his role in the systems with which he interacts, and the clear implication of his responsibility for these systems. The domination and mastery of nature is rendered unacceptable in light of the recognition of the implied domination and mastery of man.

1.7. Lessons from the Situation in Physics

The development of the argument of this chapter has contained occasional allusions to what we interpret as basic similarities between insights of modern theoretical physics and the goals of general systems research. It is also a corroborating and interesting observation that of all the disciplines physics proper has been least involved with the development of general systems research. As somewhat of a summary of this chapter we consider certain themes relevant to this issue.

An important fact which cannot be dismissed is the simple historical sequence of discipline-generation which, in the context of modern science, puts the development of physics at the beginning of the list. This temporal ordering has spared physics the pains of what economists might refer to as the imitation effect. Under this effect the natural economic growth of a particular country is burdened, hindered, and often not allowed to occur at all. This occurs as a result of perversions instigated through imitation of patterns which were suitable and right for another country which the country in question is trying to imitate. Generally, say as in the case of Latin America versus the United States, there are compelling reasons that one country 'developed' more quickly and further than others. In the case of the United States there were many such reasons, not least of which is the phenomenal abundance of natural resources. Similarly, in physics there are probably reasons for its firstness. What they are is not extremely important, but it is likely that part of the answer lies in von Foerster's Theorem Number Two:

> The hard sciences are successful because they deal with the soft problems; the soft sciences are struggling because they deal with the hard problems (von Foerster, 1972).

How much more forceful are the implications of the theorem when augmented to include the fact that because of the imitation effect the 'soft sciences' are not only dealing with the hard problems, but are dealing with them in a framework designed to solve soft problems (which, as follows, also turn out to be pretty hard).

Because of its freedom, then, to develop naturally, it appears that physics has in fact come to conclusions – even regarding the particular and more simple systems on which they focus – which are in many ways similar to those which represent the necessity for general systems research.

Without developing the associated background and context of these conclusions, it is not possible to give a complete and meaningful statement of them. Certain general principles, however, are so impressive – at least with respect to the message they represent regarding our approach to knowing – that they retain relevance even out of context. These principles are closely tied to results involving the quantum theoretic description of light (the ultimate nature of uncertainty), the theories of special and general relativity, and the extension of Heisenberg's uncertainty principle by Bohr's statement of the principle of complementarity. Holton, in consideration of the classical (before 1900) versus the quantum period, summarizes the situation as follows:

> A chief thema of the earlier period was continuity, although it existed side by side with the atomistic view... a chief thema of the more recent period was discontinuity, although it existed side by side with the wave theory... In the older physics classical causality was taken for granted, whereas in the new physics the concept of indeterminacy... as an inherent aspect of natural description [was] beginning to be accepted. In the older physics, the possibility of a sharp subject-object separation was not generally challenged; in the new physics it was seen that the subject-object coupling could be cut only in an arbitrary way (Holton, 1973, p. 118).

The implication of these dichotomies led to the crystallization represented by the 'point of view' advocated by Bohr and 'conveniently termed complementarity' by which 'evidence obtained under different conditions must be regarded as complementary in the sense that only the totality of the phenomena exhausts the possible information about the objects' (Bohr, 1949). The relatively simple phenomena which physicists had so dilligently attempted to 'understand' for centuries has led to the realization that only by considering the 'totality of the phenomena' – even allowing conflicting and 'mutually exclusive' parts of this totality as valid – can one attain comprehension of the focus of interest. How much more encumbent upon the social scientist to question the validity of simplistic descriptions of social phenomena. 'The integrity of living organisms and the characteristics of conscious individuals and human cultures present features of wholeness, the account of which implies a typically complementary mode of description' (Bohr, 1958).

We have also alluded to the fact – also attributable to a large degree to historical sequencing – that the growth of physics occurred intricately connected with the growth of mathematics. It was thus relatively easy to meet the important criteria for the justification of mathematical formalisms – indeed the question of the suitability of such formalisms has never represented a serious problem. Since the development of much of significant eighteenth century mathematics was done specifically in the context of physics, it was never forced to fit and pervert its natural inquiry and growth to pre-existing mathematical constructs. But, as we have seen, this historical development of physics was paralleled by the consequent development of mathematics purportedly separate from sense and observation-generated inquiry.

The social sciences then, in addition to the unnatural dependence on method emulation of physics, have developed without the significant power for generalization and formalization which mathematics represents.

The message of these considerations implies the impossibility of an acceptable ultimate object-generated knowledge. The interpetation which must be given to the (successful) past utilization of observer independent partial models, and of reductionist, mechanistic, and deterministic explanations, is not of their pre-eminence as a mode of description, but of the fact that with respect to certain purposes and certain themes, and in light of the acceptability of certain objectives and certain constraints, they have proven useful in the past. Where this utility has not been shown to exist, alternative conceptions with potential for meaningful results *should* be considered. Where this utility has been shown not to exist, alternative conceptions with potential for meaningful results *must* be considered.

A statement of the goals of general systems research must involve these considerations by inclusion of at least the following two objectives:

1. The articulation and dissemination of the message of these considerations as relevant to the study of complex systems within the biological, behavioral and social sciences.
2. The recognition of the relevance of this message for the study of (often unique) complex systems which do not clearly belong to any disciplinary area. (The importance of such one-shot systems is attested to by the recent boom in 'interdisciplinary studies.' That these studies have proven, on the whole, unsatisfactory is an inevitable outcome of the lack of a meaningful communicative framework.)

Any but the most superficial and cursory review will show that these message dissemination goals and thus the indirect utilization of systems research have been achieved.

An important task facing general systems research in the immediate future is the development of the means by which the 'awareness' represented by these two goals can be directly and meaningfully utilized within contexts of specific concern.

This task translates into the pragmatic goal of the development of a methodological framework which utilizes the conceptual results of general systems research. One result of the achievement of this goal is that the resultant framework would retain continual contact with and derive its basic constructs from those languages whose meaning resides primarily in extensive referrents. At the same time its generality provides the possibility for meaningful utilization of mathematical constructions, many of which have been developed without regard for their possible extrinsic meaning.

Similar considerations would apply to many of the results from computer science. Since the phenomenal growth in ability and attraction of computing machines, this area has become much like mathematics in the formalist tradition in that the machine itself, or even an abstraction of this machine, provides all the meaning that is required, in the eyes of many computer scientists.

Finally, we expect a methodological framework to serve also to integrate traditionally metaphysical considerations with research oriented toward more particular objects.

Thus, at the same time that this language traffics and interacts with languages whose meaning is primarily intensively generated, the language and the methodological framework must maintain the immediate potential for object related interpretation.

2. General Systems Methodology as Language

Per il Pragmatista, dunque, non c'e un'ipotesi metafisica che sia piu vera di un'altra.

Giovanni Papini
Pragmatismo

Facts are good, of course... give us lots of facts. Principles are good... give us plenty of principles.

William James
Pragmatism

2.1. Thematic Influences on Knowledge

One of the main points of the last chapter was the impossibility of removing all traces of a knowing subject (and the history of the subject and the state of the subject at the time of knowing). Closely related to this is the importance of themes which affect this knowing indirectly in that they pertain more properly to the cultural group and to the times in which the more localized acts of knowing take place. Recognition of this importance has come from many sources and perspectives.

Fuchs (1968), for example, describes the importance of thought patterns which are inherent in Indo-European dialects for the experience-ordering appoaches represented in Western physics, 'la nuova scienza' of Galileo; Holton (1973) traces in detail the effects of 'thematic influences' on major breakthroughs in scientific thought, showing often unexpected sources of particular 'discoveries'; Mantoux (1928) argues, even regarding specific inventions of tools and machines, that their history 'is not only that of inventors but that of collective experience, which gradually solves the problems set by collective needs' (p. 206); Schrodinger (1932), in an essay entitled *Is Science a Fashion of the Times?*, describes the necessary condition for meaningful scientific discoveries, that the community at large also be willing to incorporate the linguistic structurings that such discoveries may imply – citing as an example the 'tragic neglect' of an experiment by Grimaldi [1613-1663] which if interpreted correctly could be considered 'the first demonstration of that indeterminacy in Quantum mechanics which was formulated by Heisenberg in 1927'; Eiseley (1958, 1961) traces the effects of cultural themes and emphases on the development of the conception of evolution, as well as that of all science: 'One period, for reasons of its own, may be interested in stability, another in change. One may prefer morphology, another function. There are styles in science just as in other institutions' (Eiseley, 1961, p. 106).

The characterizations by Schrodinger and Eiseley are particularly important from our perspective in that they emphasize not only the fashionable nature of science but also the undecideable nature of certain questions. The rehabilitation of discredited viewpoints must be taken in the context of the rehabilitators, and not as representation of ultimate vindication. The situation regarding the importance of the cognitive mode of collective knowing bodies in thus analogous to that which faces the individual, and all the more relevant are self-referential limits affecting the results of cognition.

The point remains as given by Bohr, the necessity of alternative and complementary descriptive modes. This is especially relevant for the attempt to explicate the methodological language which constitutes the subject of this chapter and appears, for example, in the tradeoff which must be made between relative emphases placed on isolated parts and on consideration of the whole. Recall Pauli's observation that while we do have choice as to which insights we gain and which we lose, we do not have any choice regarding the inability to *only* gain insights.

Just as there has been overfascination with the light of the analytic method, there is equally the danger of denigration of this light due to impressive fascination with the equally (or, *also*) impressive light of alternate emphases. This expression of egocentricism is equally harmful and meaningless and precisely the danger referred to by Neumann (p. 22). Just as one can positively assert the necessity of either of two mutually exclusive modes of description, from the point of view of either description one can negatively assert the deficiencies of the other.

While there seems to have been elements of straw-man construction in some of the early and polemical general systems and cybernetic literature directed at 'classical science,' an encouraging aspect of the development of these approaches is the recognition that such negativism (or egocentric positivism) is ultimately counter-productive. Weinberg (1975), for example, after demonstrating the futility of carrying the 'strategy of reduction to its limit' makes the 'same demonstration for holistic thinking.'

Ashby, one of the earliest and most important contributors to the development of systems theory, wrote shortly before his death that:

Systems theory is essentially a demand that we treat systems as wholes, composed of related parts, between which interaction occurs to a major degree. No one supports this demand more willingly than I do but... having won our battle for the admission of interaction, we must now learn moderation...

Systems theory, having broke away succesfully from the extreme 'classic' attempt to treat the whole as consisting of isolated parts, cannot go to the other extreme (Ashby, 1972).

Ashby explicity recognized the difficulty of translating the conceptual insight of systems research into a practical formulation. His conclusion was that the exploration of this translation – and its limits – should be a primary focus of attention of those involved in systems research.

As a final statement of this recognition we quote von Foerster and explicitly point to the contrast of the following statement with that of Stigler, given at the beginning of the first chapter, in reference to great economists who must be 'accomplished in hucksterism.' Von Foerster states:

If we are after fame and success we may ignore the profundity of... problems in computation, ordering, regulation and entropy-retardation. However, since we as cyberneticians supposedly have the competence to attack them, we may set our goal above fame and success by quietly going about their solution... Competence implies responsibilities... We must share what competence we have through communication and cooperation in working together through the problems of our time. This is the only way in which we can fulfill our social and individual responsibilities as cyberneticians who should practice what they preach (von Froester, 1972).

We note that, after observing the profundity of the task facing us, von Foerster does not ask after who will solve the problems, but rather that we should use our 'competence to attack them... by going about their solution... in working together through the problems of our time.' We have referred to the importance of thematic influences of the times on the basic approach taken to scientific activity. One of the main themes of our current times appears to be that the situations of most pressing concern are in fact situations which exhibit such a multifaceted, interconnected and complex nature that it is unrealistic to expect piecemeal discipline-theoretic approaches to supply all of the necessary answers. One of the ways in which this theme has affected scientific activity has been to stress a shift, in various contexts, from paralytic concern with 'universal' theories to ones which are workable within the particular contexts and without a priori exclusion of possible influences on the situations of interst.

As we will argue below, the distinction between theory and model is not extremely important to the perspective we develop, but for our purposes here it is interesting to emphasize the return – which such a theme embodies – to the original emphases of 'science': pragmatism, relativism, and skepticism. These themes, which expressed themselves in the seventeenth and eighteenth centuries as the need to complement the all-encompassing emphases of the contemporary systems of thought, today express the need for an integrative framework to make sense of and integrate the results of largely piecemeal emphases.

One of the first pragmatic expressions of this theme is represented by the banding together of interdisciplinary teams for the study of modern problems. Weaver's justly famous advocacy of such teams in 1948 expressed their justification – in terms that became somewhat of a model for the expression of the need for systems research – by distinguishing between 'simple systems' and 'disorganizedly complex systems' on the one hand and 'organizedly complex systems' on the other (Weaver, 1948; see also Weinberg, 1975). While such teams, in principle, are an advance over isolated research, they represent merely a first reasonable access to the problems they address. Anyone who has been involved in such research is painfully aware of the difficulties created by attempting, essentially, to reconstruct the 'whole' of the object of study by putting together the parts represented by members of the 'team.' Weinberg (1975) uses a relevant analogy from the technology of large electronic devices. Such devices are constructed from a large number of individual parts, all so purely manufactured that trouble in the overall device is rarely caused by failure of one of the parts. 'On the other hand, troubles frequently arise in the joints... Why? Because the purity of the transistor is achieved by pushing out to the joints all the dirty problems.' First experiences with interdisciplinary teams only emphasize the need for a linguistic – that is, communicative – framework which meaningfully addresses the problems in the joints, and the development of this framework must be a primary goal affecting general systems research.

2.2. Methodological and Pragmatist Emphases

Implicit in the utilization of interdisciplinary teams is the shift to emphasis on problems which society faces and which current terminology might label 'large-scale.' The step beyond this, which is taken by cybernetic and general systems research, is intimately tied to the view of the dynamic nature of knowledge, and of man and society, as constituting an intricate part of these 'problems.' As such, the traditional view of a 'problem' as something demanding a 'solution' is replaced by one which emphasizes the cooperative 'working-through' of those situations and the achievement of a new and more comprehensive perspective.

In 1868 William James wrote:

> I have been growing lately to feel that a great mistake of my past life... is an impatience of results. Inexperience of life is the cause of it; results should not be too voluntarily aimed at. They are sure to float up of their own accord... and I think the work as a mere occupation ought to be the primary interest with us (James, 1868).

This view of James was to be developed in the next decade into repeated elaborations of the themes of skepticism, relativism and pragmatism, whereby James doggedly attempted to reintroduce the role and the importance of humanity in the structure of the universe. He continuously attacked the positions of the 'tough-minded rationalists and positivists' in a message which is prophetically cybernetic, taking into account both the influencing and influencable nature of 'reality' by speaking of it 'as something resisting, yet malleable.'

In his view even the notion of an absolute unmalleable world is indispensable, especially in that, as it affects the lives of those who subscribe to it, it contributes through them to changes of 'whatever in the outer order depends on them.' This theme also appears in Polanyi (1961), who argues that 'asserting and believing is an action which makes an addition to the world on which knowledge bears. So every time we acquire knowledge we enlarge the world' (p. 1). It is just in this heritage that the cybernetic and systems view, rather than accept the rationalist reality – 'ready-made and complete for all eternity' – takes the pragmatic perspective that 'reality is still in the making,' accepting the 'smallest and recentest fraction... that comes to us without the human touch' but recognizing that it has also 'to become humanized in the sense of being squared, assimilated, or in some way adapted, to the humanized mass already there' (James, 1907).

In this light, then, our focus changes from an emphasis on requirements for solving problems to one stressing our ability to find and describe problems, a focus by which we accept our role in the making and structuring of the universe.

That this pragmatic orientation is indeed an important current theme affecting scientific activity is evidenced by diverse efforts to develop frameworks from different perspectives which embody the basic aspects of this theme. An important aspect of these efforts resides in the attempt to construct a bridge from the power of abstracted and formal expressions to the reality of situations of concern which are related to humanistic systems. Just as James found that the pragmatic and humanistic perspective is 'evidently a difficult one to introduce to novices,' so too would it be difficult to understand the importance of these efforts from the perspective of either the extreme of toughminded rationalism or that of romantic subjectivity. Some important approaches, all of which are primarily directed at problems which have been 'pushed out to the joints,' are described briefly below.

1. The theory of fuzzy subsets which has been introduced by Lotfi Zadeh. Research in this area has proceeded from the fundamental insight by Zadeh that the foundational role that the framework of set theory has come to play in science and mathematics is limiting in that most pragmatic situations do not admit the ability of the knower, investigator, or researcher to unambiguously define a set. Indeed, it is possible to argue that it is just this lack of precision which is responsible for vitality and the dynamic possibility of knowledge growth (Bohr, 1963). Zadeh's orientation is an important response to Ramsey's warning that the chief danger to knowledge is excessive scholasticism, 'the essence of which is treating what is vague as if it were precise', and that 'understanding' must involve reference to 'a multitude of performances any of which may fail and require to be restored' (Ramsey, 1929).

   The need to take account of imprecision expresses itself in that the primitive notion of 'belongingness' to a set, which must be decided yes or no, is too restrictive. The extension introduced by Zadeh, to allow the notion of subset to more closely fit what appear to be the ways in which people could be said to use it, simply redefines subset so as to distinguish varying degrees of membership. This translates to a generalization of characteristic function such that its range may be determined by the practical facts of the situation which demands description. While various debates, of a scholastic nature, are possible regarding whether or not fuzzy subsets represent anything 'new,' they are basically irrelevant in that they miss the point of the paradigmatic, pragmatist, and suggestive nature of the concepts involved, a nature amply evidenced by utilizations in varied and diverse contexts. While it is possible that formally equivalent utilizations potentially existed outside of the context, the facts are that such utilizations did not exist. (See Gaines, 1976a and Gaines and Kohout, 1977 for a useful discussion of the issues, survey of relevant literature, and an extensive bibliography.)

2. Atkin has recently developed a fruitful approach to the study of structure and structural conceptions which is based on the fundamental mathematical and logical ideas of set theory and binary relations and which relates these to abstract theoretical constructions from algebraic topology, utilizing concepts such as simplicial complex to give ideational significance to practical and useful procedures. The development is especially important because of its general utility – that is, especially, because of its potential for use in the study of humanistic systems such as architectural and urban design (Atkin, 1974a), chess and modern art (Atkin, 1974b), and operative politics in the university and other social systems (Atkin, 1977).

   Atkin's work responds to the systemic emphasis on approaches which do not a priori isolate components of an overall area of study, but which take into account the total relations used to define the system, allowing components and structure to derive from the specification of the overall relations. The two most important aspects of the approach are: the utilization of mathematical concepts to allow as unambiguous as possible a specification of the basic ideas, while at the same time not pretending to the delusion of a sharp subject/object distinction, and not excluding the augmentation of results by specific circumstance interpretation; and the provision of the 'basic need of a working algorithm,' which satisfies a pragmatist orientation, emphasizing not so much the need to 'copy realities' but the need to provide the means 'to be put into such working touch with' them as to be usefully guided in the mutual adaptation of investigator and environment (James, 1907).

3. Proceeding from the working assumption that 'systems engineering of large-scale... systems can *only* be accomplished by an interdisciplinary team' Wymore (1976) has developed an overall methodology which utilizes standard engineering interpretations of systems concepts. This perspective embodies an attitude toward 'model' which – for most social scientific research – may seem useless, or even distasteful, in its emphasis on design. Although Wymore recognizes the difficulties in eliciting an adequate problem description there is an implicit belief that the overall problem can, in fact, be precisely stated. At the least, the methodology is restricted by such demands and Wymore pays little attention to such elicitations other than through utilization of 'discussions with the client' (cf. Checkland, 1978).

   This perspective appears, for example, in the use of a definition of *state* which demands that 'each state of the system must contain all the information necessary to compute the desired output of the system at any time' (p. 43). This minimally implies the knowledge of the desired output of the system at any time, and, for social science research, underestimates the importance of differences between traditional engineering-context uses of systems concepts and their utility for the study of social systems (cf. Zadeh, 1974; Florman, 1976).

Nevertheless, Wymore's attempts to extend the engineering perspective from concern merely with problems which can be specified represents a significant advance in sophistication over past systems engineering efforts (e.g., Wymore, 1967), and extensions from this base hold much promise for future research.

We also mention in this context the work of: Zeigler, which is important in its attempt to formulate an overall modelling framework through the integration of standard automata theoretic concepts, modern computer simulation techniques, and elements of general systems research (Zeigler, 1974, 1976); Pichler, which provides a general framework for the modelling of dynamical systems (Pichler, 1975), and a general framework from which to consider system decomposition (Pichler, 1978); Warfield (1976, 1973), Checkland (1972, 1975), and Gomez (1978a, 1978b), devoted to the development of methodologies which retain primary emphasis on ill-defined problem situations.

The common characteristic of each of the approaches just considered is that each exemplifies a view toward science similar to that expressed by Stevens in 1936: 'Nowadays we concede that the purpose of science is to invent workable descriptions of the universe. Workable by whom? By us.' The pragmatist emphasis thus generates a methodologically oriented ontological and epistemological viewpoint, defining what is and what we know in terms of processes of action. It is worth considering the 'scientific' status of such approaches.

An important aspect of these frameworks, and indeed most of general systems research, stems from their methodological emphasis. The point is especially well made by Bunge's characterization of general systems theories as 'empirically untestable theories' (Bunge, 1969). They are untestable in that, when considered separately from utilizations of them, they are typified by non assignment of 'factual' interpretation to most of the symbols they use. As such, they are never directly 'refutable' or 'falsifiable.' The characteristic which distinguishes them, however, from traditional abstracted frameworks is precisely the methodological emphasis, their motivation in terms of descriptions of basic activities representing what scientists actually or potentially can *do*. The result of the methodological emphasis is the continual demand for immediate extensive referability to object-oriented studies. This bestows on them, in Bunge's words, 'vicarious testability'.

We have already argued that general systems research represents a formal and communicable expression of William James' philosophical pragmatism. It is interesting to note James' belief that in the 'triumph' of pragmatism, 'science and metaphysics would come much nearer together' (James, 1907); Bunge, in fact, states that the 'existence [of general systems theories] refutes the claim that there is a sharp demarcation line between science and metaphysics.'

Bunge continues the argument that, even from the point of view of a more absolutist ontology, this clear ability to hook on semantic interpretation qualifies most of general systems research as 'scientific'. In terms of our ultimate desire to turn this back to the social sciences, it is interesting to notice that Papandreau (1958) used similar considerations to argue that economics is not a science and that what passes as 'theory' in economics (theory being what characterizes science) is really nothing more than 'model building.' Papandreau is a little less lenient than Bunge, but his concern also was with 'semantic incompleteness,' arguing that economics is not 'theoretic' since 'the class of permissible interpretations is not adequately defined'.

While we agree with Papandreau's rejection of economics as a science by more classical standards, it seems that Bunge's position regarding 'degrees' of semantic reference is as far as one can go when looking at the actual history of science. Much of the supposedly clear-cut universality of what pass as 'scientific laws' has essentially been due to the assumptions generated from long exposure to dogmatic or metaphysical realism (cf. Heisenberg, 1958).

Indeed, the pragmatist viewpoint, rather than judge the issue, merely observes that no practical significance adheres to questions of whether or not certain theories or conceptual schemes are 'correct' as mirrors of reality. The standard of appraisal 'must be, not a realistic standard of correspondence to reality, but a pragmatic standard. Concepts are language, and the purpose of concepts and language is efficacy in communication' (Quine, 1950).

2.3. General Systems Research as a Methodological Language

The linguistic turn which Quine introduces is a fruitful one, although the identification of concepts as 'language' with no further ado may be somewhat misleading. One of the most important contributions to our current understanding of the nature of knowledge has been made by the advances in our understanding of the role that language plays in the shaping of what is known. For example, we refer to the reference to Fuchs' description of the role that the Indo-European languages have played on the development of physics and to the following statement by the linguist Benjamin Whorf:

Formulation of ideas is not an independent process... We dissect nature along lines laid down by our native languages. The categories and types that we isolate from the world of phenomena we do not find there because they stare every observer in the face; on the contrary, the world is presented in a kaleidoscopic flux of impressions which has to be organized by our minds – and this means largely by the linguistic systems in our minds. We cut nature up, organize it into concepts, and ascribe significances as we do, largely because we are parties to an agreement that... is codified in the patterns of our language... This fact is very significant to modern science, for it means that no individual is free to describe nature with absolute impartiality, but is constrained to certain modes of interpretation even while he thinks himself almost free (Whorf, 1940).

The recognition of the pervasive role that language plays suggests that to best understand the framework presented below, as a response to the goal referred to at the end of the first section of this chapter, it is worthwhile to consider all of general systems research as fulfilling a linguistic – that is, conceptual and communicative – function.

In utilizing the amorphous phrase 'general systems research,' however, the important linguistic component, syntax, is left undeveloped, where by syntax we roughly understand the specification of a structure or relations holding among the fundamental concepts. It appears that lack of a communicable structure has caused unnecessary difficulties in communicating the use of general systems concepts. Evidence of this difficulty appears, for example, in the statement by Melcher that 'systems theory... puts in bold relief that traditional analyses often lead to partial analyses and misleading conclusions. Yet when one closes in on the systems concepts and attempts to apply them, their meaning and usefulness seem to evaporate' (Melcher, 1975b, p. 3). This restriction on use constitutes a restriction on meaning (or semantics) and thus presents a further restriction on the development of the language.

One of the important tasks, therefore, in the overall development of general systems research is the specification of a communicable syntax or structure. In the context of this framework structure involves a relational scheme which integrates concepts used; but further, the pragmatist nature of the framework allows a deeper enrichment by specification of this relational scheme in terms of methodological procedures, thus emphasizing operational *process* as the primary feature.

We observe that even by accepting a more linguistic and pragmatic criterion of truth than Bunge, this does not let general systems research off the hook regarding testability. The methodological emphasis only underscores the 'vicariousness' of its referential aspects, that is the lack of immediate 'real-world' meaning. Indeed it implies, as well as the ability to contribute to object-oriented studies, the ultimate *dependence* on enrichment by representative expertise which can only develop through intensive study of the object concerned. It will thus be necessarily intended to serve in a symbiotic relation with investigators from specific disciplines.

In comparing the proposed developments of general systems research with that – as we have roughly described – of mathematics in the last century or so, we can fully appreciate Castonguay's verdict on the ontological status of mathematical constructs: 'From a logical point of view, i.e., aside from heuristics, in mathematics there is no speaking ontologically, and no speaking of ontology' (Castonguay, 1972, p. 143). This may be a stronger statement than is necessary, but overall his argument is persuasive. Especially important to our purposes is his development (also referring to a similar suggestion by Bunge) that *meaning* of constructs cannot be determined either solely by the class of referrents of the construct or by the class of constructs to which the construct is syntactically related, but in fact is determined by both. Castonguay observes that although the dualistic view of meaning is actually as old as the explication of the two potentials for meaning, it is surprising that this dual dependence is so often disregarded and we find continued attempts to explicate meaning solely through one or the other aspect.

The consideration by Castonguay of the intension and extension components of meaning, and his ultimate conclusion of the need for dualistic explication, can be considered as a reflection from a focussed perspective of a slow but consistent restructuring of Western scientific thought toward the recognitition of the primary role of interaction, even to the point of opposition or contradiction, and thus to the complementary nature of different concepts, principles, descriptions, and/or viewpoints.

We have already referred to this recognition from the point of view of physics. Its expression can equally be found from other perspectives. For example consider Piaget's evidence from the study of the growth of knowledge in individuals regarding strict subjectivist or behaviorist positions, that:

> A certain equilibrium between assimilation of objects to the subject's activity, and the accommodation of this activity to the objects thus forms the point of departure of all knowledge and is presented at the very outset in the form of a complex relation between the subject and the objects, which simultaneously excludes any purely empirical or purely apriorist interpretation of the cognitive mechanism (Piaget, 1970a, p. 108).

Laswell (1961) argues that the study of political science and law requires 'systematic multivalued models of the social process.' Rosenblith (1961), after surveying some major issues of current research in neurophysiology, asks if there is 'a moral that imposes itself' and responds 'we believe that there is, and we believe that it can be stated in a single word: pluralism.'

It is impractical that this restructuring will immediately affect all of science, nor is it necessary. Perhaps it is not even desirable. For as long as there is a significant body of thought within which to incorporate narrow perspectives, even these more narrow perspectives may be useful and helpful.

As more scientists are willing to accept the importance of their work on its merits, and the contribution it can offer in conjunction with other work, rather than striving after 'fame and success,' then it becomes easier to properly evaluate and benefit from even those efforts predicated on an approach of self-aggrandizement and on the belief, as described by Stigler, that only by eschewing a 'balanced and temperate statement' and by instead 'shouting my wares... will my fellow [scientists] read on' (Stigler, 1965, p. 5). It becomes easier to recognize the insight and merit in the works of many scientists and philosophers, and to separate this from the often accompanying vulgarization and perversion of balancing and complementary viewpoints, often achieved by strained attempts to create artificial distinctions. This is especially important in trying to understand important issues related to the question of subjectivity/objectivity and to incorporate this understanding in a meaningful approach to investigation.

It is interesting, for example, to compare the non-fanatical and positive views cited above of Piaget, Ashby, von Foerster, and Weinberg, with Popper's admission that, 'Much of my work in recent years has been in defense of objectivity, attacking or counterattacking subjectivist positions.

And by an objective theory I mean a theory which is arguable... one which does not merely appeal to our subjective intuition' (Popper, 1974, p. 138). This is of course a simplification of many respectable subjectivist positions. If indeed all that Popper has been 'attacking and counter-attacking' has been theories that merely appeal to our subjective intuition, then it is unlikely that there is even any straw in this straw-man. Masterman may have more closely hit upon the rationale for much of this gladiator image of recent years in observing that 'actual scientists are now increasingly reading Kuhn instead of Popper' (Masterman, 1970).

The integrative emphasis which general systems research represents leads naturally to attempt its expression in a form with most potential for interaction with scientists from different disciplines and different perspectives. This expression should then provide a means in which to accept the position that, in situations embodying seemingly contradictory general and fundamental principles, each of which has a long and illustrious history, it is pragmatically more meaningful to recognize that, rather than impoverish our perspective by going to the wall in defense of one principle or another, it is useful to accept complementarity and recognize that more than one view may be necessary. Indeed, in many cases it goes further than just the statement that two principles are each necessary and allows that the tension which results from the opposition of the two is often more important than either of the principles (Gaines, 1978a).

Lerner has similarly stated that the 'fact that the ancient dichotomy between quality and quantity is now conceived as a creative tension between reciprocal ways of knowing bespeaks a deep transformation in the structure of Western thought' (Lerner, 1961a). The language of general systems research should reflect that transformation.

In the rest of this book we develop a general systems methodological framework, the construction of which is based on the following fundamental considerations:

1. We accept Piaget's findings that 'knowledge is never derived exclusively from sensation or perception but also from schemes of action or from operatory schemes' (Piaget, 1970a, p. 86). To constitute an internally generated contribution the framework must thus represent both categorization ('The development of formal categories is, of course, tantamount to science-making' [Bruner, et al., 1956]) and the capability for the development of operatory schemata involving the resultant categories.

2. The framework is basically intended to be able to represent the results of general systems research as a language.
3. In this regard a primary requirement is that the formulation of the fundamental concepts must indeed represent the generally accepted sense in which these concepts are used. Because of the nature of general systems research, this argument applies on two levels:
   a. The framework must represent the concepts as used in general systems research.
   b. The concepts in general systems research must in turn adequately express a meaningful sense of their actual and potential use in more interpreted languages.

   This is important because of the danger of attempts to pass off poor mathematics as systems theoretic results. General systems research can, of course, *motivate* certain mathematical research, as well as utilize and enrich it.
4. The language is intended to have special significance for utilization in social science research. Special emphasis is placed on the role of the framework as intermediary between object-oriented languages such as political science, anthropology, and sociology on the one hand and more highly abstracted languages such as mathematics, computer science, and philosophy on the other.
5. The most fundamental distinction affecting the two types of languages referred to in (4) is basically associated with differences in respective emphases on intensively and extensively generated meaning; the one putting primary emphasis on internal consistency, the other representing a high degree of semantic reference. Overriding concern in the middle-level language which we develop in succeeding chapters will be given to the necessity for both semantic and syntactical components.
6. The language is thus intended to serve the role of a *natural language* for scientific research. Because of this the framework must retain the ability for adaptation to incorporate new and/or shifting emphases which may develop. As such, its dynamic nature precludes a sharp distinction between syntax and semantics, especially with respect to the development processes. As structural (relational/transformational) adaptations occur, these provide new avenues for the development of semantic reference; as new associations develop through interaction and translation to more object-oriented languages these, in turn, directly specify or instigate further syntactical developments.
7. The intention that the language serve as a logical and structural source of

abstracted, accepted concepts and that it serve as a methodological and operational framework must be given major emphasis. In this way the framework will be able to offer external structural meaning to the object languages with which it interacts. Because of the nature of the derivation of the concepts, the framework as external structure thus represents a basis which can serve to help the object-oriented languages in the generation of internal structural meaning, a mechanism often referred to as 'theory construction' in specific disciplines.

8. The language should be flexible enough to interpret and incorporate in a meaningful way results from mathematics, computer science, and philosophy. This, of course, holds for new as well as existing results. In this way it may fulfill a significant role with respect to supplying external meaning to those studies where meaning is primarily generated from within themselves.

3. Basic Concepts

3.1. Categorization and Uncertainty

An important factor which must be considered in the development to this point is the relatively short period of time that the need for a general systems methodology has been properly articulated. A main consequence of this is that there does not exist the generally agreed to common language which is our main concern. This is of course compounded by the fact that the very nature of general systems research implies very different backgrounds and conceptual and referential frameworks of the investigators involved. The full breadth of common identifying characteristics is thus not immediately obvious. This natural situation is not in itself damaging and in fact the process of exchange and discovery among scientists from various areas has contributed much to vitality and growth. Without precluding the continued potential for such vitality, however, it is also worth guarding against the very difficulties which have motivated systems research, those stemming from the tendency toward overspecialization. This has been traditionally discipline-driven and elements have carried over to within general systems research expressing differences in emphases which stem from the original backgrounds and motivations of different investigators.

Recalling the eight points specified in the preceding chapter as underlying the development of a cohesive and operational framework, it is apparent that specification of categories, or acceptable fundamental concepts, is the first necessity in such development. Bruner, et al. (1956) note that categorization represents one of the most ubiquitous and fundamental phenomena of cognition. In the context we are employing, cognition is analogous to the ability to respond to the question of what general systems methodology is. In addition to the more obvious achievements of categorization – the reduction in the necessity for constant learning – two others which the authors list are especially relevant for the development of this framework. They are:

- the opportunity which it permits for ordering and relating classes of events; and
- the direction it implies for instrumental activity.

We thus correspondingly expect the existence of an acceptable conceptual scheme to provide important insight regarding the syntactical development of the framework and, in fact, to already and of itself contribute to the goal of supplying direction for various instances of research.

There are of course various directions that categorization attempts could take and in choosing one or another the basic purposes of the attempt must be taken into account. From our perspective, the two most important are generality and methodological utility. In each case we are again led to our primary concern: the desire to contribute to the supply of pragmatic and general theoretical concepts for use in the social sciences, in a manner which implies operatory capability without perversion of context and which, at the same time, offers the possibility of motivation and utilization of mathematical research.

With these purposes in mind it is impossible to disregard the long-standing recognition in the social sciences of a basic fact which physics has more recently come to, that of the uncertain nature of things. This has expressed itself methodologically in the pervasive utilization of probabilistic and statistical techniques. One of the difficulties facing this utilization, however, is again the problem caused by the split of mathematics from contextual reference. As statistics has developed as a branch of study in its own right; as statisticians have developed their technique more and more as technique simpliciter and less and less as technique embedded in a meaningful framework, technique which develops and grows in response to contextual demands; then all the more poignant has becaome Tukey's 'hypergeneral principal':

> While techniques are important in experimental statistics, knowing when to use them and why to use them are more important (Tukey, 1954).

Tukey's argument that good mathematical statistics need not be good experimental (or useful) statistics, and that statistics must 'continually compare its logical structures with the logical structures *currently used* and being put into use by science... and other fields' indicates that statistical considerations, too, supply a rationale for the framework we propose.

Tukey's article is provocative and indicates a breadth of knowledge and overall comprehension about methodological problems and the relations between data, data-processing, statistical theory, etc., that it is impractical to expect from more than a very few scientists. The general issues behind the problems he describes of course affect not only statisticians but workers in every field – including general systems, especially as it comes to be regarded more and more as a separate field of study.

A somewhat recent development which has important implications regarding the need to include probabilistic considerations in the construction of a methodological framework has been that of information or communication theory. The original articulation and development of this approach to uncertainty was made in connection with narrowly technological considerations and constituted a large part of the initiation and growth of cybernetics. Its fundamental conceptions, however, provide an approach to various questions for which much of statistics has been developed to answer, but it does not embody many of the hidden assumptions and linguistically camouflaged difficulties leading to one of Tukey's main worries – that the experimental statistician is too often led by this camouflage to fool both himself and his client as to what he's 'really' doing.

While the original development of information theory was closely tied to hardware transmission problems, the recognition that the fundamental principles were in fact mainly based on the ideas of transmission and communication led to the further exploration and development of these principles in the context of general systems. The original development was primarily associated with Hartley, Shannon and Wiener, and the later extensions primarily with Garner, McGill, Miller and Ashby.

The information theoretic approach which gives methodological specification to the idea of communication – and thus interaction – between two or more entities has been valuable for the development of a reference frame which appears to closely parallel 'logical structures' which are actually used in investigative situations. Not least among the relevant aspects is the fact that scientists are often involved with the utility of their descriptions and structures for the development of general theoretical statements, and thus must be concerned with nuances of their relation to them. By not making restrictive demands on the original distributions, the basic theoretical devices – information, transmission, interaction – supply analogues to the traditional correlation coefficient, variance, etc., which are less likely to hide what the actual data does and does not say.

The major formulation closely parallels the use of the concept of entropy in physics and just as entropy has had a profound thematic influence on modern physical science, so has its analogue influenced general systems research. There are various reasons for this, not least of which are the technological and methodological implications, but most significant is probably the fact that the basic conception provides an expression of uncertainty which relates fundamentally to the whole of that part of the universe which the investigator has chosen, distinguished and described for study. There are therefore pressing logical reasons for the utilization of information theoretic formulations within our framework. As we have mentioned, Ashby (simply described by von Foerster as 'that practical man') has played an important role in enriching the original formulation of the information (uncertainty, entropy) measure by developing its utility and relevance for general systems conceptions, especially through his development and elaboration of the notions of variety and constraint. With this concept – and with many others, including the basic notions of system, of regulation, self-organization, etc. – Ashby has provided pragmatic formulations which are both operational and rigorous without resorting to primitive and naive conceptualizations implied by many attempts to argue against (that is, to disregard the importance of) the purposes and effects of the investigators using these formulations. Thus, many of Ashby's formulations are particularly suited to our needs.

The temporal context of Ashby's work demanded a close attention on his part to the creative and foundational function of the crystallization of scatterings of ideas which in many cases were totally dependent on one or another of the languages existing on either ends of the extremes which we have been considering. While information and communication theory was significant especially because of its fruitful use, this was not the case with all the conceptions which Ashby transformed. The effect of Ashby's treatment of many of these, which had shown signs of existing elsewhere, was to strip away much of the special-language and dysfunctional veneer which has a tendency to form as secretion from entrapment in too narrow a perspective. Charles Peirce, the American philosopher, in arguing for the use of logic diagrams, proposed they would allow one to get the 'most difficult and esthetically as well as otherwise intellectually, iconic conception of them likely to suggest circumstances of theoretical utility, that one can obtain in any way' (in Gardner, 1958). While Ashby's approach involved a mixture of diagrams, simple examples, symbols and words – and the discarding of excesses in each of these categories – his ability to construct the guts of the issue involved makes Peirce's statement particularly appropriate for a description of his work.

Because of his recognition of the need to pull together and develop ideas from a variety of sources, Ashby, and others such as Heinz von Foerster, did not directly address the issue of the development of the relations between all of the fundamental concepts they dealt with, though there is implicit cohesion running through most of this work. Consequently, much of the framework derived here can be considered to be motivated, at least indirectly, by considerations articulated by them.

One of the overall problems which Ashby and von Foerster did address was the explicit consideration of the role of the investigator or observer in the definition of systems (Ashby, 1956; von Foerster, 1973). Explicit recognition of the importance of this role can be considered as an indirect statement of the need for alternate and complementary modes of description regarding any overall object of investigation. This emphasis points to the recognition that a nonstatic and evolutionary conception of knowledge requires interplay between various modes expressing themselves, for example, as subjective/objective, conceptual/operational, abstracted/contextual, or qualitative/quantitative and that the implications of the resultant dynamism require that static descriptions be incomplete descriptions (cf. Pattee, 1976). We would like to retain this feature in the development of this framework.

3.2. Epistemological Levels

While different methodological frameworks have been developed in recent years – for example, Wymore (1976), Mesarovic, at al. (1970, 1975), Zeigler (1976) – the particular perspectives which each of them represents, resulting of course in special strengths and in certain situations for which they are most ideally suited, also make them less suitable for the purposes described in the last chapter. The most significant work which can be related to the specified goals has been the systems methodological framework as developed by Klir. In the last decade or so Klir has developed an approach which incorporates the major expressions of the scientific process which can meaningfully and usefully be said to apply to systems from the widest spectrum of disciplines. This has resulted in an elaboration of the primary concepts fundamental to general systems research with a special emphasis placed on the descriptive mode most suitable for their transformation to pragmatic significance. This work has been particularly fruitful because of its evolution to a communicable conceptual framework which embodies a significant breadth of scientifically useful categories. This framework is primarily based on the grouping together of the fundamental general systems concepts into a description of systems defined on various epistemological levels. The overall framework developed here is based on this epistemological hierarchy as elaborated in Klir (1969, 1975, 1976, 1978c).

At this stage in the development of general systems research the approach and direction initiated by Klir is not only important but crucial if it is in fact true that there are significant situations – in particular, societal situations – for which general systems research can prove to be not just vaguely relevant, but useful. Boulding recognized such a need in 1956 in his statement that the development of a framework of coherence was 'the great task of general systems theory.' In the intervening twenty years it seems that this task has become more pressing. In the sequel we consider evidence of disillusionment in the social sciences which is attributable to the fact that much research in general systems does not, as with Tukey's statisticians, consider the difficulties faced by actual research in the disciplines. Even a friendly commentator such as Bunge detects growing skepticism due to the fact that 'GST's have not deliverd all the goods promised by their most enthusiastic proponents' (Bunge, 1977).

The approach initiated by Klir is based on the belief that general systems research does have goods to deliver and that these goods are based on the considerations made in the last chapter, not least of which are those oriented to the reintegration of the knowledge process. But this belief is augmented by the recognition that achievement of this reintegration requires more than the statement of its desirability.

The intention in the explication of the concepts which follow is that the underlying and primary consideration be methodological, subject to the constraints of generality and the ultimate intent of a cohesive and connected framework.

We use the word 'methodology' here in two ways:

1. In its etymological sense, as the study of method. That is, we conceive that a general system process description is in fact a generalization of discipline-specific systems principles. Thus parts of the procedural aspects of our framework result as a study of methods, in particular – but not limited to – scientific methods.
2. In the more common current sense, as 'a body of methods... a particular procedure or set of procedures' (Webster).

Thus, through the study of the principles or procedures of inquiry – of methods – which are used in diverse contexts, we also generate a body of methods useful within the general systems framework. None of the words method, methodological, methodology are to be considered as referring to anything other than in the sense of general systems.

3.3. Primitive Concepts

Given that the basic elements of our framework are admittedly always one step removed from the 'real world' and that it is not intended to encompass the more abstracted or meta-language of philosophy, we must take as primitive certain concepts which are necessary to the framework, but about which we can have little to say. The three most fundamental are: *investigator, environment,* and *object*. We accept that any situation of interest involves an investigator. The investigator may be an individual, a team, or even the conventional wisdom of a (conventional) group (say of political scientists or sociologists).

We also take as given that for any situation of interest to our framework the investigator is confronted by an environment with which he interacts. Through this interaction the investigator constructs, conceives of, formulates, or discovers some object upon which, for certain purposes, he is interested in performing an investigation. Certain aspects of the purposes of the investigation will ultimately be incorporated in the general systems portion of this framework, but especially at the level of object-determination they are not. The actual generation of objects, of fundamental things, is then beyond our scope but the description we accept is basically at odds with both idealism and naive realism and recognizes, along with Piaget and Quine, that at the fundamental level 'epistemology is a chapter of psychology' (Quine, 1969). We also note that this position fully recognizes the importance of the particular culture, history, and relevant extant knowledge affecting the investigator (such knowledge could, but need not, include that of the framework developed here).

3.4. System Traits

One of the basic questions which a general systems framework must be able to answer is: What is a system? A commonly given response is that a system is a set of interacting parts. While this definition probably captures the most essential aspect of system, it does not offer the full range of potential for profitable use and communicability. Three other definitions have been offered by Weinberg, Ashby and Wymore:

As any poet knows, a system is a way of looking at the world (Weinberg, 1975, p. 52).

The *system* now means, not a thing, but a list of variables (Ashby, 1956, p. 40).

A system is an assemblage $Z = (S, P, F, T, \sigma)$ such that:
1. $\sigma(f, o) = w(S)$ for every $f \in F$;
2. $\sigma(f \rightharpoonup s, t) \circ \sigma(f, s) = \sigma(f, s+t)$ for every $f \in F$, $s, t \in T$ such that $s+t \in T$;
3. $\sigma(f, r) = \sigma(g, r)$ for every $f, g \in F$ and $r \in T$ if $\mathrm{res}(f, R[o, r)) = \mathrm{res}(g, R[o, r))$ when $r \geq 0$, or if $\mathrm{res}(f, R[r, o)) = \mathrm{res}(g, R[r, o))$ when $r < 0$.

The system Z is said to be time scale determined if:

3′. $\sigma(f, r) = \sigma(g, r)$ for every $f, g \in F$ and $r \in T$ if $\mathrm{res}(f, T[o, r)) = \mathrm{res}(g, T[o, r))$ when $r \geq 0$, or if $\mathrm{res}(f, T[r, o)) = \mathrm{res}(g, T[r, o))$ when $r < 0$ (Wymore, 1976, p. 397).

To ask which of these three definitions is correct is to ask the wrong question. Each of them represents important aspects that should be involved in the definition of system and each puts different emphasis on certain modes of description; each is 'correct,' and the framework developed in the following chapters is intended to be able to incorporate the essential aspects of each of them.

As a first step toward deriving a framework involving these different aspects Klir has specified various systems traits (see Klir, 1969, 1978c), combinations or subsets of which constitute the various pragmatic conceptions of system as it can and has been used. The rationale behind this endeavor is not to categorize all aspects of scientific or general systems knowledge 'complete and ready-made for all eternity' but to offer a pragmatic, usable, and dynamic core that is intended to be augmentable, prunable, and adaptable to various needs and situations. Rather than explicitly enumerate and consider a set of traits as the set to be used in explicating our conception of systems, we do so only implicitly through the elaboration of concepts most useful for development of our framework.

In the last section we specified certain notions which will be primitive to the overall framework. Since the development of the framework embodies implicit and explicit transformational rules between concepts and groups of concepts, it is convenient and fruitful to consider the totality of the methodological general system-defining concepts as ordered according to various epistemological criteria; *epistemological* in that the framework represents knowledge regarding the object of investigation which has been primitively chosen (through the object system, as will be described), and *ordered* in that different systems concepts of proven utility often implicitly define the need for or the knowledge of various other concepts. At any rate the concepts which evolve generate definitions of systems at various levels such that for a system at a given level, the defining characteristics of that level are such that it is most often immediately possible to derive system descriptions at a lower level, whereas system descriptions at a higher level demand more information – ultimately, a structuring or organizing decision and input on the part of the knower.

3.5. Basic System Types

3.5.1. OBJECT SYSTEMS

In terms of specifiable utility the lowest level systems definition, which is directly related to the methodology we derive, is basically that as given above by Ashby. The essence of Ashby's observation that system is 'not a thing' is not that 'things' don't exist but that it is not possible to define 'thing' at the (general systems) methodological level. Recognition of this fact is given, for example, by Bunge (1974) who there 'addresses [what] the systems theorist must take for granted... categories such as those of thing.'

The characterization of the lowest level entry into the methodology is that the investigator, in recognizing a particular object as investigable, methodologically defines that object by the *listing of certain attributes which he chooses as relevant to the purposes of his investigation.* This simple characterization of a thing, which pragmatically recognizes the need of an observer (investigator) to define attributes, merely crystallizes an ancient psychological position as expressed, for example, by Kung-sun Lung of the Eastern Chou Dynasty (c. 300 B.C.), that 'Things consist of nothing but their attributes. But attributes are not attributes in and of themselves' (in deBary, 1960).

While specification of an object by attributes gives an overall sense of the object's constitution, it is inadequate from any perspective which embodies a conception of process. The problem of time and, more generally, motion and process is undoubtedly a significant problem from a purely metaphysical perspective. From our perspective we have simply to recognize that individuals know (or grow to know [Piaget, 1970b]), and science basically reflects the knowledge of motion or change in conjunction with a residue of non-change, or invariance. This change/invariance juxtaposition reflects itself in all levels of our framework and, most simply, at the current level by recognizing that the notion of methodological object is not complete without recognizing the potential for different appearances or values of the (invariant) attributes. While it is possible to identify an attribute strictly as the (name of the) set of appearances which are admitted, this becomes inordinately cumbersome, too far removed from actual usage, and conveys a sense which is basically at odds with activity and, thus, implies structurations which are not complete.

With respect to the recurring apposition of changing and invariant aspects as fundamental to process in this framework, we accept that knowledge and knowledge growth occur as a result of the juxtaposition, but recognize that invariance itself is to be understood only as relative and thus also transitory. That is, there are no absolute invariants and each invariant aspect may itself be considered as variant with respect to a different emphasis of the overall process. For example, the set of attributes, which possess at least methodological identity, is modifiable with respect to the object in terms of an overall investigation.

At the level we are currently describing we give content to an implied construction against which both change and invariance can be perceived through the specification of an additional set of attributes against which the investigator chooses to measure the attributes chosen to define the methodological object. This further emphasizes cultural and historial influences through the assumption of the existence of measurement or reading procedures through which different *appearances* (values) of the attributes can be recognized (see Pattee, 1976).

In terms of the development of this framework as a methodological complement to various other disciplines it is obvious that not only the choice of attributes but, more fundamentally, the choices of appearances and the means and capabilities for acceptable measurement must be determined from within particular disciplines, that is, from the perspective of extensive object-knowledge.

The concepts described thus far represent a sufficient means to define *system* at its most basic level. Such a system is referred to as an *object system* and thus requires the three concepts (see Klir, 1978c; Luce, 1971):

1. attributes which constitute the methodological object (*basic attributes*);
2. attributes which constitute the field against which the basic attributes are observed or measured (*supporting attributes*);
3. a set of appearances for each basic and supporting attribute.

It is often the case, though not necessary, that the sets of appearances possess some further pre-systemically determined structure, such as order relations. In such cases utilization of the following conceptions are considered to inherit that structure.

3.5.2. GENERAL IMAGE SYSTEMS

The transformation of a system, or object system, to a general system is accomplished through two steps:

1. Depending on the purposes of the investigation, measurement capabilities, and the data-processing facilities which are available, not all possible appearances of attributes will be recognized as distinct. Generally, a partition is defined on each set of appearances in consideration of these factors (or, for each attribute certain of the appearances are *lumped* [see Zeigler, 1976]), allowing that the partition could be trivial in that all appearances are recognized as distinct. This step constitutes a determination of a *resolution level* for the investigation.
2. The appearances of each attribute are generally given in terms of units related directly to the name of the attribute. Generality is attained through the assignment:
   - to each attribute of an interpretation-free symbol or name (that is, of an abstract variable);
   - to each of the lumped classes (for each set of appearances) a symbol which is also interpretation-free; these symbols are most conveniently chosen to be the positive integers.

The results of these two steps, a *general image system,* is thus a system specification in a standard form which facilitates the development of general methodological procedures. The set of symbols representing the lumped classes of a given set of appearances may also be considered as the resolution level of the associated attribute (and thus of the abstract variable chosen to represent it).

Since measurements against the set of supporting attributes often do not require distinctions with respect to individual supporting attributes, in these cases it is convenient to refer to the collection of supporting variables as the *parameter* of the investigation, and to the Cartesian product of their (general) value sets as the *parameter space.* In the following chapter a simple example is given of the transition from object to general image system and the reader may also choose to concurrently refer to Part 2 of Chapter 6 where a complete example utilizing the overall framework is developed in detail.

3.5.3. DATA SYSTEMS

The object system and general image system together constitute the *source system* for the investigation in that, while the system defined at this level is independent of observation, the general image system implicitly contains all possible dynamics, that is, all possible trajectories or mappings from the parameter space to the cartesian product of the (general) value sets of the basic, object-defining attributes. While the parameter space could consist merely of a time set, there is of course no reason to so restrict the space (that is, to restrict the supporting attribute to time) and for the cases of most interest to social scientists it is necessary that the parameter space not be so restricted. Thus *dynamics* will often be used to imply motion through social groups, cities, countries, geographical space, cultures, etc., or through combinations of these, as well as through time.

A concept which will be used at different levels is that of state. For our purposes we utilize the basic definition of *state* as given by Ashby, that of 'any well-defined condition or property that can be recognized if it occurs again' (Ashby, 1956, p. 25). At this level the well-defined properties are simply the values that the variables can take. We thus speak of state of the parameter, states of individual variables, or state of the system, the latter two usually as the state with reference to a particular value of the parameter. At higher levels the concept will be specialized somewhat to a more standard and less general usage, but we note here that our use of state need not, and in general will not (or cannot), imply or assume 'the ability to predict all future values' of the system.

As we have stated, the intention of this framework is as a methodological language capable of providing a cohesive structure which can operate in conjunction with investigations from various disciplines. Most such investigations involve a choice from among the set of all possible trajectories as given at the level of object-system (or source system). Such a choice is often the result of an empirical investigation and we consider the augmenting of an object system by such a choice to constitute a higher level system, which we simply refer to as a *data system*. Thus a data system differs from an object system merely through the definition of a function (usually through observation and data-gathering though, in the context of design, this function may be a priori given) from the parameter space to the set of possible states of the basic variables.

3.6. Higher Level Systems

3.6.1. FRAMEWORK/INVESTIGATOR INTERFACE

The transparency of the concepts defined with respect to object and data systems is important in maintaining a connection between the object – through these definitional levels – and the more methodologically interesting systems defined at higher levels. While all aspects of general systems research must implicitly embody some such mechanism, the failure to make such considerations explicit often contributes unnecessarily to the vicariousness of the results.

The major characteristics of systems above the level of data systems have primarily to do with simplification. That is, after having chosen, defined and observed the object of investigation, the primary interest resides in giving some simplified description involving the assimilation and transformation of what is known to that point, but simplified only in the sense that important information is not discarded. Of course, the determination of what is important is a major problem which can, in general, not be made without reference to the object and purposes of the investigation. Nevertheless, just as an individual derives categories and operatory schemes adaptable to various external referents, a methodological framework is predicated on the conviction that certain processes with communicable structural meaning also contain acceptable referential (that is, data) components.

In more straightforward terms the situation can be characterized as follows: The specification of a data system – attributes and observations – serves as a channel of communication from the package,

investigator, environment, object

to the set of

concepts and operatory schemes representing general systems methodological research

In an even cruder sense we recognize that these concepts and schemes are what are intended to help the investigator 'make sense out of his data.' The important difference between this set-up and a more traditional use of mathematical research is that these concepts are embedded in the overall methodological framework which is developed in the context of general systems research. In this manner data are not just abstract symbols but observational representatives. The existence of such a framework embodies an efficiency which increases the likelihood that the adaptation and development of mathematical constructs retain connections with scientific investigation. It also becomes more likely that recognizably general problems (for example, organization and regulation; this recognition ultimately must be the result of a communicative process within the community at large) receive the attention of different investigators offering different perspectives and insights. Thus it is easier to recognize those aspects of the problems which can meaningfully be abstracted and methodologically developed.

We mention at this point that the progression from the object to the concepts we are about to define need not in practice follow the strict epistemological order as presented here. It will, for example, often be the case that the investigator will be able to formulate a direct semantic reference to the higher level concepts. This will generally reflect the results of object-specific research and represents equally meaningful utilization of the framework.

It is also worthwhile to consider a more general point that should be obvious and has been made from different perspectives. Huxley, for example, observes that the beneficent function of language which grants access to the accumulated record of others' experience also involves the danger that the surface codification, by which linguistic concepts are most easily transmitted, may dupe the user of the language into mistaking the codification for the overall process or awareness which is signified (Huxley, 1954). Actually, it is useful to make the distinction between those aspects of situations which can be unambiguously communicated and those which yet require insight and an awareness of the overall patterns and relations. In this sense we use 'language' more in the manner of Whorf in that it 'embraces all symbolism, all symbolic processes, all processes of reference and of logic' (Whorf, 1942, p. 252), and of Cassirer, recognizing that neither cognition nor language 'is a mere mirror, simply reflecting images of inward or outward data' but that it 'effects a new mediation, a particular reciprocal relation between [subjective and objective] factors' (Cassirer, 1955, p. 93).

Thus, while any of the concepts and operatory schemes which become integrated into this methodological framework can be used simply and mechanically, the more meaningful interpretation is that various results be taken only as the most distinguishable components of part of an overall scheme requiring amplification and elaboration in various directions. These involve, of course, the tenuous reinterpretation to context to determine what sense the abstractions have in terms of the object. This sense should be further supplemented by amplificatory study of the conceptual and linguistic significance which underlies the more formal and abstracted operations.

Here, we may refer to Peirce's distinction between *Explicative* and *Ampliative* reasoning. Explicative aspects refer to those parts of the overall framework which are precisely formulable. These include definitional portions, but especially relate to other, generally mathematical, aspects which will be incorporated into this framework through specifications of tools for solving particular systems problems. These components are essential for the elicitation and clarification of patterns and relational aspects which exist, but generally in as complicated and hidden a form as would be impossible to determine them without complex procedures.

We recognize, however, that the results of their utilization do no more than to make explicit what is already implicit in either the data or the premises or assumptions which are used to develop the procedures – or in the mathematical system underlying them. Meaningful overall utilization of this framework, then, requires use of the results as a means to go beyond the explicative aspects through utilization of what Peirce refers to as Ampliative Reasoning and which 'includes almost all the reasoning upon which any stress is laid except by mathematicians and by formalists and pedants' (Peirce, c. 1900, p. 211).

We have observed that 'making sense' out of data generally implies simplification. The two most widespread forms of such simplification involve a determination of so-called 'laws' which the observations represent (or which represent the data), and a further determination of what it is that is behind these laws. Actually, the phrase 'behind these laws' may be misleading in that it carries connotations implying a prejudgement of fundamental issues which have not been resolved in centuries of debate, issues involving the primacy of structure versus function, synthetic versus analytic modes of knowing, induction versus deduction, theory versus data, chicken versus egg. This framework should be regarded as compatible with any position, with primary emphasis on the provision of pragmatic directions in which to proceed.

3.6.2. BEHAVIOR SYSTEMS

The next system level which we define represents simplification procedures, the main purpose of which is to free consideration of the system as much as is pragmatically possible from individual consideration of values of the background variables. That is, procedures are utilized to determine a description of the object-defining attributes which does not depend on the identity of particular states of the parameter. This can be described most simply as the determination of relations among the sets of values of the basic variables. It is often meaningful to augment knowledge of the relations – which we refer to as parameter-invariant – with a probability distribution representing frequencies of the different elements in the relation.

In the social sciences the main procedures used at this stage have almost exclusively been associated with curve-fitting and, even then, almost solely linear regression. Indeed, in many cases social science 'theory construction' is indistinguishable from determining the coefficients of a system of linear equations. While the single-method-driven nature of such attempts is invariably overdone, it is important for general systems methodology to recognize the need which such situations imply for usable ways to derive general implications from observational systems. We will consider the status of such techniques in some detail in the next chapter.

Systems at the level we are considering essentially describe the *behavior* of the object, and the relations which are determined represent a mechanism for generating data with the same statistical structure as the data-system. If the parameter space has inherited an order relation from the referents of the supporting variables (as, for example, if time is involved), the indeterminate nature of the overall behavior can often be cut down by using *translation rules.* These are described further in the next chapter, but they primarily specify that at any reference parameter value the *state of the system* is considered to include knowledge of the history of the system to that point or, more generally, knowledge of states of some basic variables at one or more other parameter values, where these other values are defined in terms of the reference parameter value. If the use of such translation rules in fact allows for a more deterministic generation of the data, then the system may be said to possess *memory*. Ashby (1956) has shown, however, that it rarely makes sense to attribute memory to the 'system,' as it is clearly a result of the interaction of the observer with the system. Klir (1975) has developed effective and straightforward procedures which further clarify and generalize these insights. The results are directly usable and, as with Ashby's formulations, cut through much nonsense which is often associated with disembodied abstraction, clearly specifying what the data (which the investigator is already a part of) can say and what has to be put in from the perspective of the purposes of the investigation.

It is also easy to show within the context of this framework that through the use of translation rules a behavioral description (that is, a partition of sequences of value appearances, augmented with their probabilities) is epistemologically equivalent to a description by system state transitions (that is, description of the movement through an ordered parameter space), requiring only a technical modification of a general procedure to determine one or the other description. This will be further described in the next chapter.

Notice that the simple definition of state is significantly less restrictive than that which is utilized in more 'classical' approaches, both classical science and classical systems theory (Zadeh, 1969; Northrup, 1947; Bridgman, 1959), in that most of the latter approaches imply the ability to know all future values or future distributions of the variables (or of output functions). Such notions have potential significance in engineering situations involving construction in the strict sense or unlimited experimental capability but are pragmatically useless for most situations of interest to the social sciences, that is, for situations which do not allow for isolation and experiment and for which the concept of 'exact replica,' which is often used as a surrogate for repeated experiment, just makes no sense.

The main issue – as with Zadeh's fuzzy sets, Atkins' Q-Analysis, and Warfield's interpretive structural modelling – is clearly accepted, and usefully so, that reality includes our description of reality. In most cases of interest to the social sciences, we 'must give up' the 'old and cherished concept' of some thing 'really' out there that fits our idea of what we'd like to be out there; we must give up a methodological orientation which sets up a predetermined image of the way things are and then considers that which we describe to be a mere approximation to this way.

The situation here is similar to that of the old view, and theory, of a 'particle', which was useful and still is in many situations. When not in these situations, however, even the view that certain information exists but is 'just not in practice obtainable' means 'clinging to the old view' which wastes energy, effort and resources in that it 'constantly drives our mind to ask for information which has obviously no significance' (Schrodinger, 1957). With respect to the social sciences we only need recall Hayek's statement that such views have not made a single contribution to our understanding of social phenomena (cf. also Gaines, 1976b).

4. Generative Systems

4.1. Introduction

In the last chapter we presented the basic concepts of the overall framework associated with preliminary aspects of an investigation. These involved the methodological primitives:

- investigator
- environment
- interaction between investigator and environment
- object.

The first system level was that of *source system,* encompassing both object system and general image system. The next level of system which was defined, that of *data system*, incorporated the ability to elaborate a particular system involving observations or pseudo-observations. This led to the recognition that natural further investigation, or higher-level knowledge regarding an object of investigation, can be expressed in terms of behavior and structure. Consideration of these two higher levels constitutes the material of this chapter and the next. In this chapter we concentrate on description and characterization of *behavior systems* and in the next we consider *structure systems* in considerable detail.

In this chapter we give an extensive consideration to problems related to the elicitation of a behavioral description and use this consideration to elaborate ideas which are especially important for the overall development of the framework. To illustrate certain fundamental issues relative to systems defined at this level, we describe here the most general form of behavior determination, based on no assumptions as to structure of the sets of values of the basic variables. Thus, the approach we describe is suitable for use with nominal, ordinal or metric variables. This involves the utilization of translation rules and, to illustrate their use, we assume that the parameter space is totally ordered (as, for example, in the case of time).

Klir's concept of *mask* (Klir, 1975, 1978c), used within this overall framework, expresses and clarifies a broad class of considerations based on the determination of behavior and explicitly generates a procedure for this determination.

One form of system description which stems from classical dynamical theory (meaningfully developed for mechanistic systems) defines a system in terms of differential equations, for example, as a set of n equations of the form,

$$\frac{dx_j}{dt}(t) = f_j(x_1(t), \ldots, x_n(t); M_1(t), \ldots, M_r(t); t)$$

where the $x_i(t)$ are state-variable functions of time, assumed given and continuously differentiable, and the $M_j(t)$ are control-variable functions of time.

Except for sporadic utilizations in mathematical economics in the tradition of Jevons' ideal, such descriptions have proven to be of limited use in the social sciences. The problem appears so obvious that it is baffling that such statements are rarely made, that being that this purported mode of description of reality is, especially with respect to research in the social sciences, linguistically and pragmatically false – pragmatically false because it's useless; linguistically false because of the impossibility of convincing anyone who is not a priori committed to the form that the form has any meaning. More specifically, there is the exhibited inability of ever coming up with a set of functions which can be taken seriously. We thus find it more realistic to work with Ashby's recognition that:

> Continuity ascribed to natural events has often been put there by the observer's imagination... Thus the real truth is that the natural system is observed at discrete points (Ashby, 1956).

One of the instruments which modern research has at its disposal is the digital computer, allowing for more massive computational capability than was available at the time that continuous mathematics developed its firm grasp on our hearts and minds. It is only very recently that science is beginning to accept that a sequence of observations need not represent an approximation to the 'true' continuous curve which must be underlying it (Greenspan, 1973) and to recognize that the observations you get are the observations you get and need not include, for example, normally distributed error terms. Many of these conceptions were developed to cope with an inability to handle significant numbers of observations.

4.2. Representation of Source Systems

Assume that an object is chosen for investigation; that a set A of n basic attributes and a set B of m supporting attributes are defined on the object; and that associated with each attribute is a set of appearances. Thus the *object system* is defined by the four entities:

1. $A = \{a_i \mid i \in I_n\}$
2. $B = \{b_j \mid j \in I_m\}$
3. $F_B = \{A_i \mid i \in I_n\}$
4. $F_S = \{B_j \mid j \in I_m\}$

where I_n is an index set such that $I_n = \{1, 2, 3, \ldots, n\}$ and A_i is the set of appearances of attribute a_i that are recognized. Similarly for I_m and B_j.

A (general) image system is determined by defining:
- a set of basic variables $V = \{v_i \mid i \in I_n\}$
- a set of supporting variables $W = \{w_j \mid j \in I_m\}$
- a family of sets $G_B = \{V_i \mid i \in I_n\}$, of values or states of the basic variables v_i
- a family of sets $G_S = \{W_j \mid j \in I_m\}$, of values or states of supporting variables w_j
- one-to-one correspondences from A to V and from B to W
- for each basic and supporting variable an onto mapping $L_i(L_j)$ from the set $A_i(B_j)$ to the set $V_i(W_j)$.

As a simple example consider an object system defined by three attributes in an attempt to investigate features of political/economic interaction. The three basic attributes chosen are:

a_1: *Rate of Inflation,* where the set of appearances could technically be defined as the set of non-negative rational numbers. Assume, however, that the investigator chooses to recognize any inflation rate below 5% as acceptable (ACC), a rate between 5% and 10% as acceptable but signifying danger (DAN), and a rate greater than 10% as unacceptable (UN). Thus $A_1 = \{$ACC, DAN, UN$\}$.

a_2: *Military Orientation,* where the set of appearances is defined as the percent of gross national product devoted to national defense outlays; considerations similar to that of inflation generate a definition of $A_2 = \{$LOW, MED, HIGH$\}$, where LOW corresponds to less than 5%, MED to appearances between 5% and 10% and HIGH to appearances greater than 10%.

a_3: *Political Control,* where the set of appearances is intended to indicate democratic or republican control over the Executive Branch, the Senate and the House; thus, each appearance is expressed as a triple such as DDD, DDR, etc., where the first component of each triple indicates a democrat or republican President and the second two represent democrat or republican majorities in the Senate and House of Representatives, respectively. Thus A_3 = {DDD, DDR, DRD, DRR, RDD, RDR, RRD, RRR}.

As supporting attribute consider that the investigation utilizes only *time* where readings will be recorded monthly for the period 1946-1975. Thus the single supporting attribute b_1 represents time and B_1 = {Jan. 1946, Feb. 1946 ... Dec. 1975}. Transformation to *general image system* is easily accomplished by renaming the attributes and the elements of the attribute sets.

Thus $I_n = \{1, 2, 3, 4\}$; $I_m = \{1\}$; $V = \{v_1, v_2, v_3, v_4\}$, where v_1 corresponds to a_1, etc.; $W = \{w_1\}$; the elements of G_B are $V_1 = \{0, 1, 2\}$, $V_2 = \{0, 1, 2\}$ and $V_3 = \{0, 1, 2, 3, 4, 5, 6, 7\}$; and the single element of G_S is $W_1 = \{0, 1, \ldots, 359\}$. The correspondences are obvious with, for example, $0 \in V_1$ representing an acceptable inflation rate, $2 \in V_2$ representing a high (greater than 10% of GNP) defense expenditure, $7 \in V_3$ representing a republican President with republican majorities in the Senate and House, and $359 \in W_1$ representing the month Dec. 1975.

At this stage the general image system so defined (epistemological level 0) implicitly contains all possible behaviors, or trajectories of states of basic variables in the parameter space, that is, all possible functions from $\times_j W_j$ to $\times_i V_i$, where W_j represent the set of all observation points and where V_i represents the set of all possible (or recognizable) observations on the defined object system. By augmenting the image system with a choice of one of these functions, say f, which is either desirable (as in the case of systems design) or a result of observations and data-gathering, we thus define a data system (epistemological level 1). In the case of the example, data-gathering associates with each element of W_1 (each month of the observation period) a three-tuple from $V_1 \times V_2 \times V_3$. Thus, for example, $f(200) = (0, 1, 0)$ indicates an acceptable rate of inflation, medium level defense expenditures and democratic everything in September 1962.

Determination of a parameter-invariant characterization of the object (a behavioral or state-transition relation) results in a system defined at epistemological level 2. Such systems are called *generative systems*, reflecting the fact that the essence of such a system is a simplified characterization which is able to generate the data of the level 1 system. As we have mentioned, the two types of generative systems are epistemologically equivalent in the sense that the same procedure, with only slight technical modification, is used to determine each one. There are, however, different connotative interpretations that can be made and different questions that can be asked which imply different perspectives from which to view the same 'facts.' These different perspectives involve different conceptual images associated with the respective interpretation and thus different theoretical implications that can be drawn with respect to the object of investigation.

4.3. Representation of Data Systems

The clearest way to represent a data system is as a matrix with rows labelled by the variables, and columns labelled by the elements of the parameter space. Thus the entries in the ith row represent the observations on attribute a_i, as mapped to states of the corresponding variable v_i, over all values of the parameter. It is possible to formalize these notions, e.g., as a projection of the image space of f onto its ith dimension, but such symbolization serves no purpose in this context, though it is useful later in considering a certain decomposition due to Ashby, that of cylindrance.

One of the most basic questions associated with what we recognize as scientific inquiry asks how the basic attributes associated with an object of investigation are related to each other (Singer, 1971). In observational terms this translates to asking which appearances of different attributes occur together in the same unit of observation. The answer to this question constitutes our basic characterization of *behavior* and is given by the assignment to each element of the product space $\times_i V_i$, i.e., to each possible observation or recognizable state of the overall system, a probability based on its frequency of appearance in the data array. Many relevant questions relate only to whether or not the probability is zero (that is, whether or not the sample appears), while others depend on the actual frequency distribution. General systems methodology can only characterize the questions and point to ways in which the information provided by various processings may be useful. It will always be for a particular investigation in terms of a particular interpretation to determine the actual relevance.

A common extension of the simple question of the occurrence of appearances of different attributes in the same observational unit is the consideration of the relevance, at any given moment, of past (or future) values of certain attributes. More generally stated, we ask whether or not, with respect to appearances at a particular observational unit, it is relevant to consider appearances at other observational units. In considering relevance, we refer to pragmatic relevance in determining respective weights to assign opposing perspectives which emphasize either the interconnectedness of all states or the need to consider each thing or state only as unique. Our major consideration is to provide a means for determining the relative emphasis which should be placed on one or another consideration in the context of the purposes of specific investigations.

From the methodological point of view we need only recognize that such determination of memory (or, in general, 'neighborhood') effects are agreed to be important in many contexts in the social sciences, for example as: lagged effects of consumption on investment, or of a change in the money supply on the rate of inflation; temporal and spatial diffusion patterns of pottery design forms in archaeological studies; neighborhood effects on the behavior of individuals, etc. It is therefore meaningful to provide methodological procedures related to such questions. Since our activity oriented perspective ultimately depends on such procedures we will describe this particular 'problem' in some detail.

4.4. Definition of Mask

Translation rules (see Klir, 1977) express these inter-parameter-state dependencies in the following way, and it is useful to define these in the most general manner possible. We do so in this section, though the general symbolic relational expression is not completely necessary for the most useful implementations. These are considered in greater detail in the following sections.

Each translation rule t_r identifies a value (or values) of the parameter – for each reference value of the parameter – which is to be associated with that value (or, if the parameter is time, a set of past or future values which is to be considered relevant for each value). Each t_r may thus be considered as a binary relation on $\times_{j} W_j$. Denote the set of all such translation tules by T.

Choice of a 'best' behavioral description is based on a choice of 'proper' parameter values to associate with reference parameter values, and these may be different for each basic variable. For example, if the parameter is time it may be relevant to consider two lagged values of a given variable and none of another. Recalling that V is the set of basic variables we express this choice as the association of translation rules with each variable, and call such a choice a *mask*. If we denote a mask by M, then:

$$M \subset V \times T$$

We call the set of all translation rules associated with a subset of V (which is often a singleton set) a *submask*. For example, M_{ij} is the submask associated with basic variables v_i and v_j if:

$$M_{ij} \subset M$$

and if all $(v_i, t_r) \in M$ and all $(v_j, t_r) \in M$ belong to M_{ij}.

It is convenient to view the states of variable v_i, either at the reference parameter value or at other translated parameter values as determined by submask M_i, as states of new variables, referred to as *sampling variables.* Thus a mask, and therefore a set of sampling variables, represents a sampling scheme which can be used for 'making sense out of the data,' that is, for deriving a simplified description of the data which more or less captures major features of the data.

A choice of 'best' mask may, for example, be used to answer the question: Which past values of which variables are most relevant to the ability to determine the current states of all the variables (that is, the current state of the system)? The issues are most easily stated in terms of time, but can easily be taken in terms of any ordered parameter-space. If the parameter space is unordered, the considerations are somewhat different and meaningful translation rules will depend on the interpretation of the supporting variables.

In most uncontrived situations of interest to the social sciences there will of course be some sense of connection or relevance for all possible sampling variables. Similarly, it will often be the case that there will not be one single most important sampling variable. In most pragmatic situations, the choice of a parsimonius description – of a simple behavioral description – involves a tradeoff on the part of the investigator between this simplicity and the completeness of the description. This statement implies that there is, in general, no 'best' choice of a mask, no 'best' selection of sampling variables in an absolute sense, but only 'best' in terms of purposes of the investigation, and tradeoffs which the investigator is willing to make.

These tradeoffs operationally boil down to the number of sampling variables involved in the description versus the amount of 'certainty' one would like to have about the current state of the system. We observe that it is of course possible to make stipulations beforehand which constrain the behavior determination; for example, that a priori demands of simplicity allow utilization of only a fixed number of sampling variables.

4.5. Choice of Best Mask, Sampling Scheme and Behavior System

4.5.1. GENERAL CONSIDERATIONS

The simplest way to describe a general procedure for determining the best mask is to consider that at each parameter state each element of the data array is a potential state of a sampling variable. The context of the investigation will suggest, relative to each parameter state, certain sampling variables to be of primary interest, in the sense that the uncertainty regarding these variables should be minimized. The choice of mask involves the selection, from among the other potential sampling variables of those whose values give most information about the sampling variables of primary interest. From the context-independent perspective of general systems there is no a priori way to determine the most meaningful partition of sampling variables into those of primary interest and those which are to be considered as 'giving information'. The case described involving the partition of sampling variables into those constituting the current values of the basic variables on the one hand and all past values of the basic variables on the other is the most natural and common.

We point out, though, that the generality of the procedure we are describing is such that the very decision regarding the original partition could be the major issue regarding which the *investigator would choose* to make use of the general systems framework. A simple example would be a situation which considered only current states of basic variables and which involved limited capacity for immediate observation. The procedure would be used to determine which of the basic variables that the history of the system indicates should receive the attention of immediate observation.

From the point of view of the data array, the general problem is defined, then, by these steps:

1. Determine which sampling variables are to be given consideration;
2. Determine a partitition of the potential sampling variables of interest into two classes, X and Y;
3. Determine a subset of X such that some criterion relevant to Y is best satisfied.

Step (1) can only be determined by context-dependent information; Step

(2), as we have observed, may be context-determined or not; Step (3) is often a problem which can be solved independently of context, given certain requirements.

Subsets X and Y are often called input and output, but with respect to situations of the most general appearance in the social sciences such terms are misleadingly suggestive. The terms independent and dependent variables are also frequently used, but these too have developed misleading connotations. We recall that the primary goal of any procedures incorporated into the framework is to provide perspectives on observations such that interpretations from the point of view of the object and the investigation are minimally prejudiced outside of context. From this view, the terms 'explained' and 'explaining variables' seem least suggestive.

While it is conceivable that cases arise which – at Step (1) – do not include all reference-state values of the basic variables as sampling variables, such cases are of minimal interest and we assume that current values of all basic variables are considered, that is that no current values are a priori excluded. We also assume that future values are not included in the set of considered sampling variables.

Assume that the investigator has chosen as explained sampling variables current states of certain basic variables and allows all past values of the basic variables corresponding to the explained variables and all current and past values of other basic variables as potential explaining variables.

The situation which this case describes is analogous to classical and well developed systems theoretical problems dealing with systems identification (Gaines, 1978b). The theory related to these problems, however, has developed almost solely in the context of machines or direct abstractions from them. While the theory and techniques have been extremely successful in developing methods for hardware construction and for describing such construction, they generally embody assumptions – for example: determinateness, ability for repeated experiment, tractable distribution functions (e.g., in computer simulation) – which leave them little or no utility for the social sciences. These systems theoretic directions which retain a heavy emphasis on the older and stricter definition of machine – and such directions are also retained in some cybernetic approaches – will continue to be useful in mostly technological situations where it is sufficient to remain at the level of performance imitation. For the social, behavioral and biological sciences, however, it is necessary to extend the narrow machine interpretation of the concepts involved (cf. Sommerhoff, 1974).

4.5.2. APPROXIMATION VERSUS COMPLEXITY

The main difficulty stems from the fact that the details of situations of interest involve a level of complexity that must embody choices regarding simplification. By predetermining a modelling form – say that of linear automata – the modelling form itself makes many important decisions which can most meaningfully be made in the context of the interpretation. Once major interest or attention is attached to a particular tool or modelling form, then the decisions and assumptions which it embodies are usually not given the attention necessary to evaluate the utility or dis-utility of the form. Gaines, for example, has shown the 'nonsensical' nature of results of modelling attempts in terms of deterministic finite state automata 'in the face of even a slight trace of acausal behaviour.' Gaines' demonstrations can be taken to show that insistence on deterministic behavioral descriptions of the indeterminate realities social scientists are interested in would lead to models which are essentially as complex as the observations they are attempting to simplify (Gaines, 1976b, 1977).

This is not to say that meaningful uses of such forms are impossible, but that such use demands that each potentially usable form be subjected to evaluation which can only be made by potential users. A major effect of prejudicial attachment to certain forms is that unnecessary special language jargon (whether purposely or innocently perpetrated) serves mainly the function of obscuring the main issues involved, thus avoiding (or saving such issues from) evaluation by those capable of making such evaluation in the context of interpreted research situations. Such avoidance, however, also has the unfortunate consequence that many uses that could be gained – though often more modest than originally assumed – are not taken advantage of.

Bunge gives an interesting caricature of the methodological situation by observing that in general systems research the emphasis has shifted from a statement that a model is the wrong model to a statement that the model is all right, it's the interpreted situation which is wrong (Bunge, 1969). This is not contradictory to the basic position of this framework, but with the important stipulation that it is necessary to develop models or forms which significant numbers of situations do not, in fact, 'flunk,' or, at the least, to augment these models with the procedures necessary for an evaluative interpretation.

The development of a general systems framework thus requires the ability to integrate both the abstracted power of various forms at all levels and a description of the ways in which this power is abrogated when meaningfully interpreted to research situations. Recognition of these requirements leads to the need, with respect to each concept of interest, for the most general description which still retains the potential for methodological utility.

4.5.3. SYSTEM IDENTIFICATION

In the problem currently under consideration, for example, the issues can be given a very general expression through use of the conceptual apparatus associated with communication theory, which provides a characterization of the probabilistic and statistical structure in terms of the overall process under description. Thus, presuming the set Y of explained variables is given, utilization of simple computational procedures determines X', a subset of X, such that the conditional uncertainty of Y over all of the process, given the values of X', is minimized. Ideally, a set of explaining sampling variables could be found such that the uncertainty in the explained variables is zero, that is, such that the explained variables are unambiguously determined by the explaining variables. This is a close analogue to the more common understanding of the notion of state, but experience with observational sequences from the social sciences supports the statement that insistence on such a set of explaining variables is likely to lead either to the nonsensical complexity that Gaines has described or to significant problem-modification to fit the model.

As should be apparent, the current behavioral description could easily be adapted to a perspective from which the primary methodological problem is described in terms of generating a state-transition description of the system. Thus, for example, the set Y of explained variables could consist of the values of all the basic variables at what may be considered as the 'next' time instant. The set of all current and past values of each basic variable would constitute the set of potential explaining variables. If a set of explaining variables could be found such that there is no uncertainty as to the values of all basic variables at the next parameter state, then the system would be said to be state determined. In terms commonly used in the context of machine synthesis, the approach as described here is similar to utilization of a finite-memory paradigm for the generation of a state-transition structure which realizes an observational sequence (though in many engineering situations the sequence is determined or given by a 'client'). In these situations, models using the finite-memory approach are often augmented or replaced by so-called finite-state methods (see Klir, 1972c; Klir and Marin, 1970) which introduce arbitrary internal variables (thus, arbitrary states) rather than sticking to the semantically defined basic variables. These arbitrary states can also serve to code – in addition to patterns of past values of variables – patterns which exist within these patterns. The performance-imitation nature of such constructions is perfectly acceptable in an engineering context, but from the point of view of the social sciences – and an objective of 'making sense' out of observations – it doesn't make much sense to satisfy the objective through sole reliance on new elements which themselves don't make sense.

Such procedures give acceptable and meaningful results in machine-construction problems since the history of the development has gone hand-in-hand with the existence of structural components, elements, or subsystems (described in the next chapter), and with developments of techniques of synthesis which use these elements. Thus, the 'sense' of arbitrary-state dynamics is inextricably connected with a structure system which supplies the sense. In the more general contexts which we are considering here the sense-giving set of elements is far from apparently given. One of the important ways in which the overall framework we describe contributes to the methodological significance of individual techniques is by introducing the behavior-structure connection to broader classes of phenomena, thus creating the potential for greater use of these methodological procedures developed from an engineering context. The extraction of insights and perspectives on the function-structure relation, and amplification of the general aspects of particular techniques is quite different and significantly more involved when taken out of very special contexts, and appears to be one of the most important directions for general systems research.

Machine realization theory is quite well developed for deterministic situations and there has been recent progress on probabilistic automata, but the development is not yet significant enough to be of much use to the more complex situations of interest to the social sciences. Even at the current stage of development, however, it is clear that the computational complexity involved in procedures used to generate models – although the models themselves are potentially less complex than models which result from assumptions of determinateness – will require input and interaction with the investigator and utilization of knowledge of the object which he may possess.

Models which introduce arbitrary states even in this context still face the problem of useful and meaningful interpretation of the states so introduced. Again, it seems that many of the difficulties involved will necessitate the development of procedures which retain connections with lower and higher level systems as described in our framework, rather than developing algorithms which work independently of the built-up knowledge.

A compromise procedure which appears to offer the most promise takes advantage of object-related knowledge and generates results which retain connection with the object. This procedure involves utilization of a combined finite-memory and finite-state approach (Klir and Marin, 1970). The method proceeds by first establishing the best sampling scheme, through use of the mask as described, followed then either by use of suitable clustering algorithms applied to the resultant states to determine the generation of new states or by the direct introduction of new uninterpreted sampling variables. In this manner the model which is derived takes the fullest advantage that is pragmatically acceptable of simplifications directly related to context and proceeds, from the results at this stage, to recode the residual uncertainty to supply methodological meaning to this residual. These new states, which are analogous to arbitrary states introduced through algorithmic procedures, thus have the dual advantage of more immediate interpretability and of overcoming some of the inherent complexity involved in most investigations.

For the rest of this chapter we will consider behavior or state-transition descriptions which define states solely in terms of sampling variables or other means allowing direct translation to the object of investigation.

4.5.4. RELATIONS TO THE SOCIAL SCIENCES

Two commonly used techniques in the social sciences which are methodologically similar in results to the masking procedure, as described, and which fit into our overall scheme of obtaining a behavioral description from observations, are those of finite Markov chains and of linear regression. Each of the techniques possesses an important similar relation to the overall framework we propose in that they are highly developed techniques, the use of which must take into account the severity of the assumptions necessary to generate specialized results associated with the techniques.

In the case of Markov chains the relation to the general procedure we have described is immediate. Let Q be a set of recognizable, logically possible outcomes. A sequence of choices or specifications from among Q, for which each choice can be considered to be based on a probability distribution function defined on Q, is called a Markov chain if the relevant probability distribution at any point in the sequence is uniquely determined solely by the specification made at the immediately preceding point in the sequence. Once the conditions of the definition are met there are many interesting results which apply to the process over large sequences of choices.

A fact that becomes obvious from even a cursory review of an extensive literature on applications of Markov chain analysis in the social sciences (e.g., Gordon and Gleser, 1974; Hoffman, 1971; Stewman, 1976; Wilkenfeld, 1974; Goodman, 1962) is that the data that the world gives an investigator are almost never strictly Markovian. In such cases, then, the possible precision associated with analytical results (such as the unique probability fixed point vector) based on the Markovian assumption is meaningless and interpretations obviously must take this into account. This does not exclude the likelihood that certain statements can be pragmatically significant in the sense of pointing to general patterns of information which can be useful in helping the investigator to understand the system he is dealing with.

If we accept this last proposition then it is meaningful to ask whether or not a process is Markovian. One statistical approach which has been devised to answer this question is based on our hypothesis-testing heritage. In this development the concept of a Markov process is extended to consider sequences of choices in which the determination of the probability distribution at any point in the sequence depends not solely on the choice at the immediately preceding point but also on the choice immediately preceding that. Such a process is called second-order Markov. These considerations similarly extend to third-order, etc.

The standard approach then becomes a test of the hypothesis that the process is of order one against the alternative that it is of order two, using goodness-of-fit statistics. These are generally derived as functions defined on the differences between the data you get and the data you 'would have gotten' if everything was going according to the abstract model. Asymptotic approximations to the 'real' distributions of these statistics then theoretically give the investigator enough information to answer questions such as: out of a hundred times that I find a data-analysis like the one I'm now dealing with, and if my hypothesis is correct, how many times will my results diverge from the true values as much as they do? While emphasis on the precision of the answers is often overdone, the tests do give an idea and feel for how reasonable a hypothesis is and how much store the investigator should put in it in his overall investigation. In the current situation this concerns the trust he should put in a particular behavioral description. As we have observed, there is in general no perfect such description and proper use of statistical techniques should be seen as input to the overall, and always incomplete, process character of knowledge acquisition--input contributing to the necessary interplay of context-related and abstracted, of observational and formal, elements.

In this regard, the masking procedure we have described is especially useful in that it gives a means to generalize the definition of Markov process and an alternative method of evaluating different hypotheses regarding order. If we consider observations on a system as recorded and arranged in a data-array, questions regarding the order of Markov chains translate to the determination of whether or not the inclusion (as conditioning explaining variables) of sampling variables located at further distances from the reference state decrease the conditional uncertainty in the explained variables.

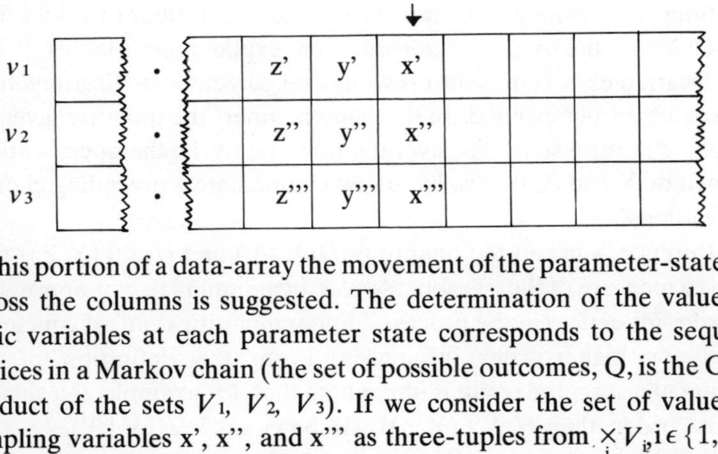

In this portion of a data-array the movement of the parameter-state marker across the columns is suggested. The determination of the values of the basic variables at each parameter state corresponds to the sequence of choices in a Markov chain (the set of possible outcomes, Q, is the Cartesian product of the sets V_1, V_2, V_3). If we consider the set of values of the sampling variables x', x'', and x''' as three-tuples from $\times_i V_i, i \in \{1,2,3\}$ and denote the set by X (similarly for y', y'', and y''' and z', z'', and z''', using Y and Z), then if the process is first-order Markov, the probability distribution appropriate to the choice from X is the same, given knowledge of Y and Z, as if just the values of Y are given. Here, 'appropriate probability distribution' assumes that we accept summary results of observations as the distributions.[1] With this structure the average amount

1. This simple common-sense acceptance is standard procedure, though usually arrived at through a sequence of arguments which renames the probabilities as 'maximum-likelihood estimates'.

of information, which is given by the specifications or choices, over the whole data-array cannot decrease as the number of explaining sampling variables increases.

Thus, the 'test' of the Markovian assumption translates to a more easily specifiable condition based on the information or average uncertainty formula:

$$H(X) = -\Sigma\, p(x_i) \log_2 p(x_i)$$

The quantity $H(X)$ represents the average amount of uncertainty represented in the choices from among X over the process (or sequence of observations; $p(x_i)$ is the calculated probability of explained-variable sample x_i).[2] Similarly, the quantity

$$H(X \mid Y) = -\sum_i \sum_j p\,(x_i, y_j) \log_2 p(x_i|y_j)$$

represents the average amount of uncertainty in the sequence of specifications from among X, conditioned by considerations of values from among set Y. In the current interpretation explaining-variables Y and explained-variables X represent previous and current specifications from the Markov chain perspective. In the same manner, the quantity given by $H\ (X \mid (Y,\ Z))$ represents the average uncertainty in the specifications conditioned by Y and Z, that is, by the two immediately preceding choices or specifications.

The difference between the quantities $H(X \mid Y)$ and $H(X \mid (Y,Z))$ thus represents a measure of the 'reasonableness' of assuming first as opposed to second order influences in the process. Similar quantities can, of course, be calculated for higher-order influences. From the definition of the information measure, the result is immediate that, for example, if a chain is indeed first order, then $H(X \mid (Y, Z))$ will be equal to $H(X \mid Y)$.

Consider as a transparent example the sequence of observations on a single two-valued variable:

0 0 1 1 0 0 1 1 0 0 1 1 0 0 1 1 0 0 1 1 0 0 1 1 . . .

2. An in-depth and pedagogically oriented development as well as utilizations of the uncertainty formula are contained in articles by Krippendorf and by Broekstra in *Recent Developments in Systems Methodology for Social Science Research,* the second volume of this series.

In this case the average uncertainty (over the whole process) associated with a single state, with no consideration of past values, is equal to

$$p(x=0)\log_2\frac{1}{p(x=0)} + p(x=1)\log_2\frac{1}{p(x=1)} = \frac{1}{2} + \frac{1}{2} = 1.$$

It is easy to see that, when looking only at current values and considering none of the past, uncertainty is maximum with either 0 or 1 being equally likely. By considering a single past value – that is, considering the sequence to be generated by a first order Markov process – it is also obvious that uncertainty is maximum. That is, the process is equally likely to generate a 0 or a 1 following either a 0 or a 1. Thus, $H(X \mid Y)$ is also equal to 1. Considering two past states, however, the process becomes completely determinate in that 00 is always followed by 1, 01 is always followed by 1, etc. Thus $H(X \mid (Y, Z)) = 0$.

Examples which give such clean results are of course rare and serve almost solely to illustrate the uses of formulae. In general, the measure can be expected to decrease slowly with a hopefully discriminant jump associated with the inclusion of some sampling variables.

From the hypothesis-testing perspective, and at the simple level of a test of one order versus another, an argument can be made that the standard statistical goodness-of-fit approaches (see Billingsley, 1961; Anderson and Goodman, 1957; Chatfield, 1973) are as meaningful as the utilization of the uncertainty measure (though Chatfield states that even at this level the interpretation utilizing the conditional uncertainty leads 'to a greater understanding'). There are, however, three major advantages which directly follow from the operatory scheme we have described.

First is the fact that the scheme immediately generalizes to allow a more inclusive definition of 'order' and eliminates the need for elaborate schemes to test interactions involving more than one variable. For example, a particular investigation could naturally lead to a chain 'order' represented by the following diagram (for greater detail regarding the procedure see Klir, 1975):

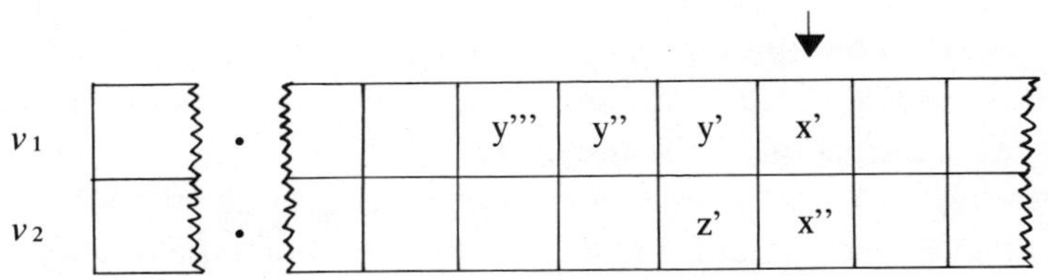

The diagram conveys the idea that three past values of basic variable v_1 and one of v_2 give the necessary information to determine the distribution function of the pair (x', x''). This is an efficiency in explanation over using three past values of both variables, and with the availability of high-speed computational facilities, it is relatively simple to write a general program which examines all past values back to a specified distance and which computes the conditional uncertainty (see Klir, 1975).

When we say 'relatively simple,' however, 'relative' is an important word. This problem underscores our main theme and an overall argument which is basic to this framework which emphasizes the utility of team research and the necessity of a cohesive and integrative framework within which the team can work. In this case the program is simple from the perspective of the abstract-level language of computer science and for someone whose major orientation and expertise is directed toward this language. It is not, however, simple for someone without this expertise. It is important to recognize that within the context of the overall methodological framework we propose here it is much more feasible to communicate the need and desire for computational capability to that component whose expertise lies in the abstracted-level language. (There is no a priori reason that expertise from different languages could not be represented in the same 'person' but the sheer quantity and extent of approaches and concepts which can and should be integrated within the overall framework makes it unlikely to retain this property over a broad enough spectrum.)

We re-emphasize also that our view toward the abstracted level languages is that, with proper communication channels to research from within object-oriented languages, new and significant results which are generated from within the abstracted language often will have immediate general systems interpretations, and thus object oriented interpretations. In this regard it is also important to reconsider that, although our framework uses mathematical and computational techniques and methods (we should also include abstractions, formalizations and philosophical principles), it provides more than just a means to integrate them in the sense of connecting them to utilizations. The overall evolutionary and process-oriented position regarding knowledge recognizes that particular tools as well as conceptions serve in a single and unified, but still distinct, manner to create overall additions to knowledge, and thus to contribute to the actual construction of the universe.

Returning to the Markov chain example, the second major advantage to the scheme using uncertainty, and embedded in a computationally feasible procedure, is closely related to the first, regarding the generation of an efficient acceptable explanation.

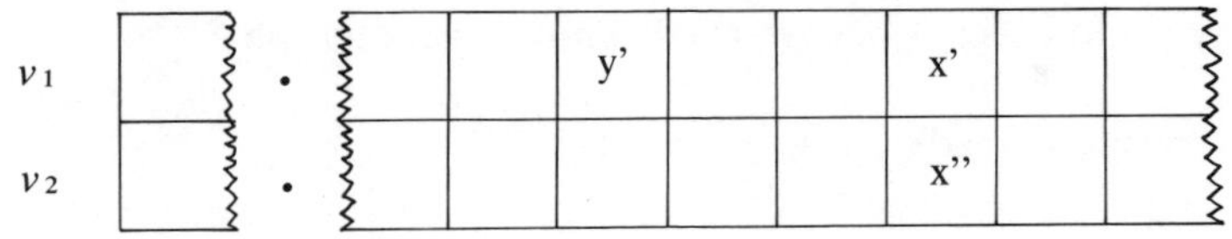

The diagram depicts a situation in which the value of one basic variable three time periods back is all that is relevant to the distribution function associated with the specification of current values. Through utilization of the computational scheme just described, this sampling variable is easily discovered whereas standard statistical tests would in all likelihood stop after checking the first versus second order assumption and would include, both incorrectly and insufficiently, the immediately preceding state. Just as it is unlikely to fall upon a meaningful first order Markov chain, however, it is also not expected to find a single sampling variable, as in this case, which will perfectly capture the patterns in the data.

This leads to the most important advantage of both the utilization of this scheme and the conception as part of an overall investigative process involving the levels as we are describing. As is implicit in many current viewpoints regarding the significance tests, and as Chatfield and Lemon (1970) explicitly recognize, the main value of any particular use of such tests is to serve as integrated heuristics for use by investigators. For example, Goodman and Kruskal (1954) state that: 'Our major theme is that the measures of association . . . should not be blindly chosen because of tradition and convention only . . . but should be constructed in a manner having operational meaning within the context of the particular problem' (p. 733).

Essential to the determination of the Markov chain order in this context, as well as to the utilization of any 'numbers' and any mathematical model, are the implications which it or they have in the overall framework. While such a point should be obvious, there is still a significant remnant of the mechanistic attitude which expresses itself in an implicit belief, for example, that the available data must have a certain form – say that of a first or second order Markov chain – and that what is necessary is only to determine which order is the 'correct' one.

Such an attitude is identical to that described by Schrodinger and referred to earlier which 'constantly drives our mind to ask for information which obviously has no significance.' Wittgenstein (1953) also comments on this attitude: 'Here it happens that our thinking plays us a queer trick . . . we encounter this queer argument also in other regions of philosophy . . . that is to say: "God sees – but we don't know"' (paragraph 352).

4.5.5. PRAGMATISM AND UNCERTAINTY

The major orientation of the general systems framework is to shift attention from the limited emphasis of the fact that reductionist models have proven successful to an emphasis on the process that leads to those successful models. Rather than search for analogues of the 'laws of Newton,' we provide a framework to discover representations of observations which lead to insights, insights which (secondary to the *fact* of the insight) then may or may not be expressable in the form that Newton expressed his particular insights.

Through utilization of the behavioral description scheme described here the investigator has the option to choose from among various behavioral descriptions of different degrees of complexity versus uncertainty with respect to the purposes of investigation. Obviously, the procedure can always lead to a 'determinininistic' situation by extending the set of sampling variables to include all of the observations. This extreme of course contributes nothing to parsimony of description but, if it is necessary, it does make the formal statement that within the constraints of the investigator's current knowledge the demand for a deterministic (non-statistical) explanation results in a complete history of the system. In the usual cases, however, this demand is not made and the investigator can make step-by-step decisions as to when the tradeoff between number of samples and simplification presents an insight relevant to his investigation.

In these examples the determinations are closely tied to the view that we have toward probabilities and thus to the problem of what it means to say, for example, that a particular sequence of observations is this or that close to being a Markov chain. The most relevant fact is that a particular data-processing is based on measurements (readings) of events which have happened, and does not imply contigency on the future. This brings up the fundamental problem of induction, and from our perspective the only reasonable approach one can take is a pragmatic one, that is, that the detection of certain patterns in the record of observations is a worthwhile activity if it contributes to the overall relation between investigator and object (or, more properly, investigator and his environment as represented in the object). In this manner we can feel that the very determination of certain quantities (such as information measures) increases awareness and thus gives a heightened sense of symbiosis between investigator and object and thus an increased likelihood of harmonious and beneficial action.

We agree with Keynes that probabilities must 'have reference to action and must be a loose way of expressing the propriety of *acting*' (Keynes, 1921, p. 339), but it is not necessary to take the restrictive viewpoint associating each probabilistic determination with a particular action. Thus, inductive determinations essentially serve as 'habits' tending to proscribe general and loose rules or directions for action (see Ramsey, 1931; Peirce, 1878). The particular implementations of these rules, and thus the actions and results they lead to, can never be determined outside of the moment of the action, and in context of the circumstances prevailing at the moment. With this less prescriptive view of the purposes of knowledge generation the scheme (or masking procedure) which we have described generates a spectrum of probabilistic statements which must be evaluated or assimilated in conjunction with the simplification each statement allows.

The point of empirical inquiry and assimilation through various processings of an investigation is not to *know* in the sense that 'nothing more need follow on that climax of your rational destiny' (James, 1907, p. 200), but to know in the sense of helping 'to summarize . . . parts of our experience . . . and getting about among them' (p. 58). If we again look more generally at the procedure we have been describing, taking it out of the stricter Markov chain determination then this most general process – as well as other operatory schemes used to derive behavioral descriptions – must be considered within this 'summarizing' interpretation. This summarizing aspect described by James is also a motivating factor in Zadeh's developments of fuzzy sets (see Zadeh, 1974). Zadeh's notion of a fuzzy variable is basically the same as the interpretation which this framework gives to such potentially precise concepts as 'order' in a Markov chain.

Though we have emphasized the more general masking procedure in consideration of the generation of behavioral descriptions, it is understood that alternate methods such as linear regression are also encompassed by the methodological framework and represent different ways to obtain behavioral (relational) descriptions regarding interaction among variables. Resort to such highly presumptive techniques, however, should only be made in the most guarded manner. The use of linear regression follows similarly to the highly developed forms more generally characterized by linear systems theory – uses which have become so highly developed not so much because they are meaningfully useful but simply because their development is tractable. While there is no reason that the use of such forms could not also provide insight, if properly viewed as (somewhat restrictive) reference frames through which to interface with observations or represented reality, the nature of social science utilizations of these forms often turns the emphasis around completely, from a scientific investigation to the fitting of investigation-related data to, say, a linear regression model (as, for example, in Hilton, 1976).

Consider, for example, McKelvey and Zevioina (1975), where the authors present an alternative to linear regression useful with ordinal variables. Their conclusion is that: 'Regression is inadequate not because of the failure to model the true relation. Rather, we assume that the true relation *is* described by a linear model, and that the failure of the regression model to describe the *observed* data is due to the inherent loss of information . . .' [author's emphases]. In the *Tractatus*, Wittgenstein wrote: 'The general form of propositions is: this is how things are.' Commenting on this in *Philosophical Investigations I*: 'That is the kind of proposition that one repeats to oneself countless times. One thinks that one is tracing the outline of the thing's nature over and over again, and one is merely tracing round the frame through which we look at it' (Wittgenstein, 1953, p. 114).

The expenditure of energy and effort on linear regression techniques in the social sciences also serves to shift the emphasis from knowledge growth as an interactive process to the mere static determination of fit to a form. The non-significance of much of the information for which these assumptions constantly drive our minds to ask is not of course unrecognized by all social scientists. But for these social scientists, not previously committed to the use of particular techniques, the recognition of their abuse often leads to such disdain toward the exhibited naivete that the disdain effectively extends to all utilizations of formalization and quantification, unnecessarily isolating such researchers from beneficial insights which can be associated with efficient means of observation characterization, and from the catalyzing inputs to creative insights that the abstracted-level languages can provide.

We re-emphasize that, in the context of this general systems framework, we do not consider formal and precise models as mere abstractions from reality, but as important complementary perspectives which can serve in an organic and dynamic manner for the overall study of an object.

At this point it is interesting to ask after the reasons for recurring commitments to particular forms which unnecessarily restrict the dynamic aspects of knowledge. The problem has a long heritage and was described, for example, by Goethe:

> This very preference for the use of formulae gradually makes the formula the principle object. An action which ought to be carried out for a certain aim becomes itself the aim, and no kind of purpose is fulfilled (in Fuchs, 1967).

We suggest that the inordinate attention to the details of such formulae, without an acceptance of their just-only complementary nature, is a continued reflection of a refusal to accept the incomplete and evolving nature of the structure of knowledge. Closely related to the orderliness and simplicity of such formulae (regardless of how symbolically complicated a formula may appear – or, as described by Wittgenstein, linguistically bewitching – the fact of its description means that it is known, and thus simple) is the promise for knowing, that is, *really* knowing, which they imply. The desire for such mastery is a residual of the scientistic attitude and of the misinterpretation of science which, from the vantage point of hindsight, chooses to place the major emphasis on certain relatively finished products rather than on processes which lead to their development, an attitude which has been described by Piaget as 'epistemological cheating' (Piaget, 1968, p. 13).

As we have tried to argue, the nature of this form of 'knowing' and will to power, while often personally and transiently ego-gratifying, has negative implications for the overall status of man, stemming from the assumption of the ability to maintain the strict subject/object distinction – an assumption flatly contradicted by modern research in linguistics, mathematics, logic and physics. (See Whorf, 1956; Heisenberg, 1958; Brown, 1969; Nagel and Newman, 1959; Josephson, 1975; Wittgenstein, 1953, 1956; Bohr, 1963; Ramsey, 1931.)

In the attempt to define a general systems methodological framework which incorporates results of modern research into the processes of inquiry, we have described levels of object characterization and systems definition which recognize object systems, general image systems, data systems, and behavioral or state-described generative systems. Just as the primary purpose of behavioral determination is to summarize and describe data systems in a simpler manner, structure systems are intended to develop simpler characterizations of the relations representing a generative system. This stage is fundamental to the methodological emphasis of general systems research, which emphasizes the need to develop techniques and approaches which proceed from descriptions of an overall perceived object of interest to a description in terms of subsystems or elements.

These subsystems are analogous to our idea of 'parts' which go together to make the 'whole', but the shifted emphasis from the search for and study of these parts prior to knowledge of the whole is a major difference in systems research from the analytic method. This distinction between analysis and synthesis turns out to be methodologically quite straightforward and, in this framework, any methodological utilization can be characterized as either an analysis or synthesis. It is important, however, to recognize that in the overall epistemological process oriented to both the construction and the description of the known it is hardly possible (or necessary) to decide which is prior.

5. Structure Systems

5.1. General Considerations

The influences which have affected the development of cybernetics and the general systems movement have also played an important role in science and thought which is less connected to a technological orientation. These similar developments, involving the recognition of the importance of structure, are reflected in the especially European movement of structuralism, described by Piaget as 'one of the most general trends of avant-garde movements in all the human sciences.' Piaget defines the position of structuralism as 'relational, that is to say as positing systems of interactions or transformations as the primary reality' (Piaget, 1970c). Broekman recognizes structuralism as an outgrowth of deep and widespread cultural trends in Russia, France, Italy, Czechoslovakia, and Poland, but argues against its being considered a philosophical current, a school of thought, or even a movement, and characterizes it simply as an intellectual orientation and as *activity* based upon common emphasis on structure. He also recognizes, however, that, 'In the writings of structuralist authors there is no consensus about what structure actually is' (Broekman, 1971, p. 7).

Louis Dumont, the distinguished French social anthropologist, asserts that 'the notion of structure is the major event of our times in social anthropology and sociology . . . After a long period dominated by a tendency which led to atomization, the essential problem for contemporary thought is to rediscover the meaning of wholes or systems, and structure provides the only logical form as yet available to this end' (Dumont, 1966, p. 41). Especially interesting is Dumont's emphasis on the need for a method based on the investigator's relation to the phenomenon he is studying. Thus the primary impression one gets from these studies is the importance of the development of methodological techniques which recognize the importance of influences not explainable by what had heretofore been considered the 'natural' elements or parts.

The clearest structuralist expression is probably represented in the mathematical structures associated especially with the Bourbaki school. Structural explication has naturally been easier in an abstract mathematical context since incumbency of semantic reference is never a primary question. Thus, structure can be observed in its purest, though most distant, form. (It is interesting to speculate that the shift to category theoretic formulations with a concomitant re-emphasis (or re-re-emphasis) of function (morphism) indicates that formal consideration of structure has reached a current pragmatic limit.)

Although the structuralist orientation gives an important re-emphasis on the role of observer as structurer, there has been a tendency to emphasize the role of relations as totally determining the identity of elements involved in the structure. This orientation, while acceptable for a purely abstract structure, has unacceptable implications in attempts to utilize structural concepts in scientific, and especially human science, contexts. Piaget has addressed this issue and asks, 'Should we, then, end our account with the proclamation 'Everything is structure' and let it go at that? No, for though it is true that everything can *become* structured, the difference in modality is all important' (Piaget, 1968). Many of the difficulties are again associated with the recurring fact of the inability of a meaningful language or linguistic mode to be consistently free from paradox or, similarly, to not run into an apparently universal conflict between, on the one hand, ideas which the very form of the language generates as meaningful expressions and, on the other, limits to the ability of the language to express these ideas.

This recognition is expressed in our framework through the recognition of continuing interaction between object-language and the general system constructions. This necessitates both complementary descriptions and potential for reformulation of the boundaries of the investigation. Thus, at any stage of utilization of this framework, particular insights may be catalyzed which demand redefinition of the object as originally conceived and re-entry into the framework with a new, but nevertheless related, investigation. Similar recognitions are expressed in the cybernetic distinction between so-called first order cybernetics and second order cybernetics; between the cybernetics of observed systems and the cybernetics of observing systems (see von Foerster, 1974; Pask, 1969).

The intention of this framework is to supply methodologically useful conceptions of structure which are embedded in the methodological framework. For any investigation, and thus with respect to any object, the framework is in turn part of the larger interactive result of the investigator/framework coupling. The importance of this coupling re-emphasizes a dynamic, incomplete, and process-oriented view of knowledge and the framework we develop supplies its essentially constructive nature with various proposed abstract structures constituting means by which to carry out the process. The role of abstract structures in this framework is thus very similar to the role which Piaget describes for them in the overall process of intelligence whereby the 'idea of a formal system of abstract structures is thereby transformed into that of the construction of a never completed whole, limits of formalization constituting the grounds for incompleteness . . . incompleteness being a necessary consequence of the fact that there is no terminal or absolute form because any content is form relative to some inferior content and any form the content for some higher form' (Piaget, 1968, p. 140).

In our view of structure and of knower as structurer with certain purposes, we do not consider structure to be strictly a property of observations, nor of the observing subject. In terms of behavioral (or state) defined systems, we are interested in determining a description (or system definition) which, given the behavior or mode of transformation of system states, can explain or reproduce behavior, but in terms of systems which are in some sense simpler or more known or more easily knowable. Thus structure is said to constitute the organizing knowledge about an object. The question arises as to whether structures so determined 'belong' to objects or whether structural forms are known a priori and are of a nature such that 'real-world' data are adapted to fit these forms. In terms of the systems-methodological definitions the answer is neither, and we recognize that structure is not strictly a property of an object, nor of a behavior system, but is a property of a system under description, and that such descriptions can never be independent of the purposes of the description.

In reference to an object and an object-language investigation our framework recognizes the need to develop operatory schemes which are context-independent in use, but whose results demand reinterpretation to context and adaptation from that perspective. It is often the case that the operatory schemes have been themselves derived from context but this is of less importance than the fact that they are recognized as useful. With regard to social science considerations, ideas associated with structure are closely related to conceptions of theorizing and 'theory construction'. As determined from the point of view of general systems, structure cannot be considered as supplying theory to discipline and object investigations, but it may be perceived as offering a pragmatist point of departure for theoretical considerations.

5.2. Definition of Structure

To generate a definition of structure with operational significance, emphasis is placed on the derivation of a description in terms of elements and relations among the elements as a mode of explanation through which observed characteristics of the overall system (behavior or dynamics) can be considered to occur. The sense of overall system is supplied in context of an overall investigation utilizing the framework through behavior systems, the subject of the last chapter. While systems defined at the behavior level can be considered to express the basic interaction among data and variables or to give a basic characterization to transformations of data which the system effects, structure systems – the subject of this chapter – offer an explanation or constructive definition of means by which these interactions occur. The notion of structure is thus related to the degree to which one or another ordering principle can be considered as applying to the object of investigation. The basic components of any explication of any given structure are thus:

1. the characteristics to which the idea or organization applies;
2. the way in which the determined structure represents these characteristics.

Part 2 of this description obviously implies that the determination of structure cannot be independent of the focus of a reseracher's interest as represented, at the least, in the characteristics chosen, and to which the structure will relate. We are immediately aware that it is not meaningful to attempt a general definition of structure which does not entail the potential for adaptability when supplied with contextual enrichment. Thus, it is reasonable for Broekman to not have found a consensus among structuralist writers as to the precise definition of structure, given that ideas of structure for most of the writers concerned derived from different orientations. In our general systems framework attempts are oriented toward capturing and specifying as much of the common elements as possible.

As we have mentioned, structure is mainly associated with ideas of organization or order. It is impossible to separate the notion of organization from the representational scheme involved, but regardless of which scheme is used the view toward order will be closely related to Huygens' observation in 1698 that:

'Tis plain then that Nature has not exhibited the Variety in her works that she could.

Thus, to the extent that structure or order is perceived in a system it is expressed as the difference between the maximum potential Variety that a description allows and the Variety which is actually observed.

Given the ubiquitous nature of probabilistic descriptions, it is important to utilize a general quantitative statement of this expression which involves such description. This is supplied through the use of entropy or information related formulas. Thus, an almost direct translation of Huygens' observation expressed in terms of modern probalistic reference frames is the measure of the amount of order defined by $H_m - H_x$, that is, as the difference between maximum entropy or uncertainty (H_m) that a description allows (which is a rough equivalent of total disorder) and the entropy or uncertainty that is actually observed or exhibited (H_x). We again utilize the entropy measure which is associated with a set augmented by a probability distribution as defined in Chapter 4:

$$H(X) = -\Sigma\, p(x_i) \log p(x_i)$$

The utility of this conception has been exhibited in communication engineering using Shannon's formulation which normalizes the quantity and defines *redundancy* in a set of messages, or a language, as:

$$\frac{H_m - H_x}{H_m} = 1 - \frac{H_x}{H_m}$$

Redundancy thus measures the extent to which the original language is superfluous (or from the point of view of message transmission, inefficient); 'The redundancy is the measure of the extent to which it is possible to compress [the language] if the best possible code is used' (Shannon, 1951, p. 124; see also Shannon and Weaver, 1949). The potential for use of this measure when viewed from a slightly generalized perspective has been noted and used by many authors (see Ashby, 1962, 1965a; von Foerster, 1960; McCulloch, 1960; Klir, 1970b; Owens, 1975) as well as in the evaluation of various behavior descriptions (utilizing conditional entropies) as described in Chapter 4.

In these conceptions redundancy is interpreted as the amount of organization that the system exhibits. To the extent that organization exists, then an observer – with a limited capacity for the amount of information or variety he is able to process and thus satisfactorily interact with – is able to take advantage of this organization for 'getting about' with respect to the organized system. Ashby in particular has explored the similarities between the message-oriented conception of redundancy and the more observation-oriented concept of constraint or 'natural law' in the sense of Huygens' expression.

Ashby's analyses of redundancy and constraint supply clearly stated basic insights into the structure of systems, and describe in detail the importance of the notion of constraint for an observer who would interact with a system (see also Porter, 1976, who explores the potential role of these insights with respect to standard control theory approaches).

In the next chapter we demonstrate that the perspective provided by systemic and organizational considerations even at this basic level can generate useful information and understanding regarding investigations by social scientists.

A notion which is closely related to the idea of constraint, and which is currently considered pertinent to most fields of investigation, is that of complexity. As we have described, one of the most important insights of modern science is the recognition that meaningful systems are in large part dependent on our descriptions of those systems – that is, on the language which is utilized. A natural concomitant of these results which is central to the framework we are developing is that problems are not 'just problems' but 'problems under descriptions' (Nurmi, 1974), and that the ability to describe or define what we consider as problems is the more significant and important aspect of the growth of knowledge. In this regard complexity is not solely an objective property of 'real-systems' which we must learn to cope with, but is equally a property of the description or conceptual scheme being used to represent the object. As the capacity for conceptualization grows, that is, as descriptive facility and methodological language becomes more sophisticated, then so does the scope of investigations – of the questions we ask of a system – become relatively more complex.

A similar theme has recently been elaborated by Rosen (1977) in the context of observables of an object. Rosen's argument is also related to our recognition that object systems can pragmatically only be considerd to be chosen with respect to the purposes of an investigation, and his basic point is that the 'relativity of system descriptions [i.e. object systems] together with the multiplicity of descriptions which provide the very definition of complexity, preclude any one class of subsystems [set of object systems] . . . from being universally valid' (p. 230).

Rosen's argument is essentially that, even if it made sense to consider the concept of a 'complete set' of observables or attributes on an object, it is not feasible to expect an investigation to deal with this complete set. Thus, in any investigation, the set of observables actually chosen represents only a relative description with respect to the object. Rosen argues that an object becomes more complex – or from a different perspective, becomes a richer object of study – as we develop more ways to effectively interact with it.

Although Rosen is mainly defining complexity at the level of the transition, object → object system, for our general system framework it is methodologically necessary to recognize different uses of the word and to define complexity at each of the system definition levels (cf. Cavallo and Islam, 1976). Rosen's major point, however, is especially pertinent to our definition of a structure system.

In the context of the overall interactive and process-oriented nature of this framework, it is necessary to distinguish two aspects of both structure and complexity.

1. From a general perspective an object is more or less complex – and it is relevant to speak of one or another 'structure' – dependent on the descriptive or representational frames which are available to the investigator. Thus, we recognize a more conceptual or *global structure* system which is dependent on the relation between the investigator (observer) and the object, and on the set of representational forms which the investigator is aware of and which can be considered to be relevant to the object under investigation.

   From the point of view of the operational component of the general systems methodological framework, it makes most sense to consider those representational forms which can be given precise (i.e. mathematical) formulation – though we recognize that they are not exhaustive – and especially to consider the most general of these. It is convenient also to only consider representations at the level of generative systems (i.e. behavior or state defined).

   One of the functions of this framework is to provide a scheme which can incorporate various representational frames, to investigate and extend their relevance for different situations, and to make these available to investigators who would use the framework. Some relevant representational schemes are thus: n-ary relations, where n is the number of basic attributes; m-ary relations augmented by probability distributions, where m is the number of sampling variables; linear sequential machines; systems of differential or difference equations; directed or undirected graphs, etc.

2. The more common use of the concept of structure is integrally related to the sense of a whole (or system) being constructed of parts, but from the point of view that – with respect to the notion of structure – the parts are less *determining* of the overall system as they are *determined* by the organization, or, the relations which hold in the overall system. As we have mentioned, the notion achieves its clearest expression in a purely abstracted context where the elements of a set are in effect defined by the 'structure' or rules of operation which define the structure.

   In less abstracted contexts, the investigator is more often faced with a situation such as that described by the physicist de Broglie:

   > As, on the one hand, there is no completely isolated particle and as, on the other hand, the bonding of the particles into a system is practically never syfficiently complete for something of their individuality not to remain, it can be seen that reality sems in general to be somewhere between the concept of autonomous individuality and the concept of a completely fused system (In Dumont, 1966).

   Acceptance of this position is implied in this framework through the recognition of the importance of the connections between the various epistemological systems levels. To translate this acceptance into operational significance we reserve the term *structure* or *structure system* as understood to apply to a particular representational scheme.

We relate structure in this sense to the 'familiar proposition that the task of science is to make use of the world's redundancy to describe that world simply' (Simon, 1962, p. 111), by defining a structure system as a set of 'elements,' 'parts' or 'subsystems,' along with a set of couplings or connections between these parts, and an implicit or explicit rule representing a means for composing these parts to retrieve the overall system. The derivation of a structure system from a behavioral system may thus be considered to be a decomposition, and we use the word in this general and looser sense to refer to any overall process possessing these characteristics. The set from which the subsystems can be chosen, as well as the rule for their composition, will obviously depend on the particular representation which is used for the behavior system.

In context of this dual conception of structure the universe thus becomes richer (or more complex) as more powerful ways and techniques to interact with systems are developed. Studies directed toward the understanding of this process are properly the foci of all the traditional disciplines, in particular, psychology, linguistics, mathematics, anthropology and philosophy, as well as general systems research. By the very definition of increasing complexity, however, this increase itself involves – in the second sense of the use of structure – the means to 'simplify' objects of investigation. The decompositions so affected may be given various interpretations, the most common being to increase our understanding of an object, thus also increasing our potential for effective interaction, by allowing a focus of attention on the simpler parts rather than the more intricate behavior system. [1]

We observe that, even given a particular representation, there is in general not *a* structure or structure system which is *the correct* representation of an object. That is, the decomposition of a particular representation using a particular technique is most often not unique. As in the choice of representation, as in the choice of decomposition technique, so the choice of one or another structure or decomposition will be intimately tied to the purposes of the investigation. In terms of the methodological framework which we are developing, recognition of these purposes will be made through the integration, into a particular structuring action, of objectives and constraints which the investigator imposes or supplies to the framework.

It is also necessary to recognize that, in addition to the fact that more than one structure system adequately expresses the characteristics chosen as invariants which the structure system should preserve, the choice of invariants will itself contribute to the acceptability of one or another structural description. In the ideal case a structure system will allow perfect retrieval of the behavior system (or, perfect retrieval of the system characteristics chosen as invariants). For most systems of interest to the social sciences, however (that is, for the most useful representational forms), we often are not able to decompose the system, even allowing for couplings between elements, into simpler parts such that the overall system is perfectly reproducible. We accept this limitation, however, and in fact incorporate it into our notion of structure. A general example illustrating

1. A somewhat common use of the term decomposition carries the sense of totally isolable parts. We consider such cases as extremes and do not restrict structure determination in this way.

these principles is developed in the sections which follow and, in the next chapter, these principles are applied in a context-related example.

5.3. Structural Derivations

The notion of complexity, in the sense of the agreement to deal with high levels of perceived interaction among entities and attributes, has especially been a concern of importance to investigators of social phenomena (e.g., Durkheim, 1893; Pareto, 1916). One of the major effects that systems conceptions and methodology have had on the social sciences and on decision-making research has been to clarify and develop ways in which the power of abstracted and operational methods can be utilized in dealing with this complexity. Structural considerations – that is, the derivation of acceptable structure systems – are of fundamental importance in that they relate the power of these methods to specific research while preserving the fundamental relation of the abstracted systems to the object of investigation.

One of the main insights of modern research into the modelling process has been to recognize the paradoxical situation which accrues to the attempt (and the need) to deal with situations and 'problems' which are not solely of the investigator's choosing, that is, to deal with aspects of reality which do not possess the laboratory and experimental characteristics of the problems traditionally chosen by classical science. The important contribution of general systems research has been to give a balancing emphasis to contextual problem-situations and to develop operatory simplification schemata which detect and express the degree to which the overall complexity of the investigator-defined object may be considered as consisting of simpler parts or subsystems.

The most general and useful considerations of structure relate primarily to the notion of constraint. If we are considering a system as a relation, then structure or organization implies restriction on the range of variety that the system exhibits. The main issues can be given a precise expression by considering a data system, defined though observation, as a function $f: \times_j W_j \rightarrow \times_k S_k$, that is, as a mapping from the parameter space to the space of recognizable data points, where S_k represents the value set corresponding to sampling variable s_k as defined in the section on generative systems.

We consider here two simple representations of behavior systems:

1. Consider that only (and all) current values of basic variables serve as sampling variables and define the behavior system simply as the image set of f. This can be symbolically described by:

$$\text{Imf} = \{d_i \in \times_k S_k | \exists w \in \times_j W_j \text{ such that } f(w) = d_i\} \tag{1}$$

Thus, the behavior system consists merely of those interactions or set of samples which have occurred. In the example from the previous chapter, if we consider a sampling scheme using one past value of inflation and only current values of the other two variables, then at any given reference parameter value (at any given month) the relevant sample consists of a four-tuple–past and current inflation rates, current military expenditure, and current political controls. In all, there are thus 216 possible samples which could occur, i.e. $\# \times_k S_k = 216$.

2. Assign a weight to each element (or data point), d_i, of Imf, signified by $p(d_i)$, such that:

$$p(d_i) = \frac{\# f^{-1}(d_i)}{\# \times_j W_j} \tag{2}$$

where # signifies cardinality. The set of such weights thus serves as a frequency-defined probability distribution over the data points. For the present discussion we consider only data systems in which:

$$\# \times_j W_j > \# \times_k S_k,$$

that is, systems for which the number of observation points is larger than the number of logically possible samples. Situations for which this condition does not hold are considered in the next chapter. Constraint is said to exist in the first case if:

$$\# \times_k S_k > \# \text{Imf}, \tag{3}$$

that is, if there are possible samples or interactions which do not occur.

In the second case, where sample occurrences are augmented by probabilities, we may determine the amount of constraint through use of the entropy measure as previously described. In a system exhibiting no constraint the occurrence of samples will be equally distributed over all possibilities and the entropy will be equal to the maximum possible entropy.[2]

In each of the two cases we have described, we consider the derivation of subsystems or parts in terms of the set of basic attributes which define the system. Thus, we allow as parts subsystems defined as behavior systems and which involve one or more of the basic variables. Thus each potential subsystem can be identified by an element of $P(V)$, the power set of $V = \{v_i \mid i \in I_n\}$ representing the basic variables of that subsystem.

To indicate the generality of the framework we are developing and its ability to integrate independently derived results, we first develop a constraint and decomposition consideration which is based on Ashby (1965b). The technique is especially useful for investigations of interest to social scientists. Its use in an overall investigation utilizing the framework we propose is a particularly clear example of the role which this framework – that is, which general systems research – serves with respect to the theory/data interface in the social sciences.

In this first example we use only current values of basic variables and admit as potentially acceptable structure systems, E_s, sets of elements, and associate with each of these elements a behavior, that is a set of occurring samples considering only the variables associated with that element or subsystem. Since we utilize only current values, then an overall sample consists of a single element from each of the value sets of the basic variables, that is, Imf $\subseteq \times_{i \in I} V_i$, where I is the index set for the basic variables. For each subset e_i of V denote the set of indices of the variables in e_i by I_i. Thus $I_i \subseteq I$. We define each E_s as follows:

$$E_s = \{(e_i, b_i) \mid e_i \in P(V); \cup e_i = V; \forall j,k, e_j \subseteq e_k \Rightarrow j = k; b_i \subset \times_{l \in I_i} V_l\} \qquad (4)$$

Following Ashby, we first add the restriction that each E_s (actually the projection of E_s onto its first component) contains all and only those elements of $P(V)$ which have the same cardinality, that is,

2. If the number of samples is significantly larger than the number of possibilities this maximum entropy will be very close to -log $\# \mathrm{X} S_k$, but for small sample sizes in which $\# \mathrm{X} S_k$ does not divide $\# \mathrm{X} W_j$ it is possible to formulate a more precise measure of the maximum entropy:

$$H_m = (p - n \bmod p) \times \frac{[n/p]-1}{n} \times \log \frac{n}{[n/p]-1} + n \bmod p \times \frac{[n/p]}{n} \times \log \frac{n}{[n/p]}$$

where $n = \# \times_j W_j$ = number of samples; $p = \# \times_k S_k$ = logically possible number of states; $[x]$ is the smallest integer greater than x.

$$((e_i,b_i) \in E_s \text{ and } (e_j, b_j) \in E_s) \Rightarrow \#e_i = \#e_j \tag{5}$$

and:

$$((e_i,b_i) \in E_s \text{ and } \#e_j = \#e_i) \Rightarrow (e_j,b_j) \in E_s \tag{6}$$

Thus, for this interpretation of structure systems, all subsystems involve the same number of variables.

The interpretation of each structure system is that the elements within each E_s represent interaction between c variables (thus c attributes) where $c = \#e_i$. For this example we can thus write $E_2, E_3, \ldots, E_n$ where $n = \#V$, and where the subscripts denote the number of elements in each subsystem of the decomposition. The criterion for the acceptability of E_s as a structure system is that the original behavior system, Imf, is exactly reconstructable from E_s. The specific definition of reconstructability is defined through the use of the operators **Pr** and **Sp**, where $\mathbf{Pr}_i$ (Imf) denotes the projection of Imf (consequently, of each observed overall sample) onto the c coordinates represented by e_i. In this way we determine the behavior b_i which is to be associated with e_i in the structure system. Thus:

$$b_i = \{x \in \underset{l \in I_i}{\times} V_l \mid x \in \mathbf{Pr}_i(\text{Imf})\} \tag{7}$$

Conversely, $\mathbf{Sp}(b_i)$ 'spreads' the behavior b_i back out over the coordinates of Imf not represented in e_i. Thus:

$$\mathbf{Sp}(b_i) = \{x \in \underset{l \in I_i}{\times} V_i \mid \mathbf{Pr}_i(\{x\}) \subseteq b_i\} \tag{8}$$

as shown in the following example:

Let $V = \{v_1, v_2, v_3\}$, $V_1 = V_2 = V_3 = \{M, F\}$,
$\text{Imf} = \{(M, M, F), (M, F, M)\}$, $e_i = \{v_1, v_2\}$.
Then $b_i = \{(M, M), (M, F)\}$ and
$\mathbf{Sp}(b_i) = \{(M, M, M), (M, M, F), (M, F, F), (F, F, F)\}$.

Recalling that constraint constitutes a restriction on the full variety possible, then $\mathbf{Pr}_i$ (Imf) indicates whether or not the interaction of the variables e_i exhibits constraint.

A potential structure system E_c is considered to be an acceptable structure system if:

$$\bigcap_{b_i \in E_c} \mathbf{Sp}(b_i) = \text{Imf}. \tag{9}$$

Thus an acceptable structure system is one for which the overall behavior system may be precisely determined by consideration at any one time of at most c of the attributes. That is, the overall system is completely reconstructable from consideration only of the smaller subsystems. Ashby has shown that if E_c is an acceptable structure, then E_d is also acceptable for all $d > c$. The minimum c such that E_c is acceptable is called the *cylindrance* of the overall system Imf.

The information which such a decomposition implies is important both from the point of view of simply being able to understand the processes and means which give rise to the perceived behavior and from the point of view of determining possibilities and limits for effective interaction with the system. It is intimately connected with considerations similar to those described earlier in the context of redundancy in the sense that as the communication engineer would be interested in the redundancy of a set of messages to be able to determine an efficient coding of the set, so would, say, a policy-making body be interested in the redundancy or constraint present in a set of messages – behavioral occurrences – to be able to derive the most efficient policy responses. If we consider the basic variables as representing system indicators to which policy-making bodies must respond, then the cylindrance measures the minimum order of the relations upon which the attention must focus in any given situation.

As in Ashby (1965b), we can construct the overall relation hypothesized by each decomposition, that is, the sequence of cylindrical closures, observing the effects as the order of relations in postulated structure increases. Observation of the manner in which increasing orders affects the approach of the postulated relation to the actual overall relation thus allows a local specification of the 'relation's intrinsic complexities.'

The information supplied by the decomposition thus can be useful even in cases where the cylindrance is not less than n (cases where the overall system cannot be reconstructed from consideration only of smaller subsystems) in that it is possible to characterize, for each value, the system states, or interactions among the defining attributes, which are responsible for the system's complexity – that is, which prevent the system from being considered as composed of the simpler sub-relations (projections). While variations in the numbers of such samples at each level represent an obvious quantitative significance, it is likely that the individual states will also possess an interpretive significance which will be dependent on the object system that is under investigation.

The cylindrance determination is especially important when questions of the distribution of samples are not as significant as the fact of the occurrence of individual states, or when sample sizes are small and probability distributions do not carry significance.

We note, however, that if the sample size is significant then the non-consideration of distributions may be misleading. For example, consider a system with three basic variables v_1, v_2, v_3, the current values of which also serve as the only sampling variables, and where $V_1 = V_2 = V_3 = \{0,1\}$. If in a large sampling all eight possible 3-tuples of the form (v_1, v_2, v_3) appear, then there would be no exhibited constraint between any pair of variables (or among all three as a unit) and the judgement would be that the overall system was totally unstructured. In terms of cylindrance this would technically be cylindrance 1, but Ashby correctly implies that such cases may equally be considered as irreducibly complex. 'Structure means the absence of certain possible configurations, and chaos means the presence of all possible configurations' (Watanabe, 1969).

If, however, the frequency distribution associated with the states is as shown in Figure 1, this judgement would misrepresent the complexity for many purposes.

v_1	v_2	v_3	$p(v_1, v_2, v_3)$
0	0	0	.01
0	0	1	.24
0	1	0	.24
0	1	1	.01
1	0	0	.01
1	0	1	.24
1	1	0	.24
1	1	1	.01

Figure 1

If we look at the projections onto two-dimensional spaces as shown in Figure 2 (these constitute the behaviors with augmented probability distributions of all subsystems which involve only two variables), we see that consideration of the probability distribution indicates significant constraint existing between variables v_2 and v_3 and none between v_1 and v_2 or between v_1 and v_3.

v_1	v_2	$p(v_1, v_2)$	v_2	v_3	$p(v_2, v_3)$	v_1	v_3	$p(v_1, v_3)$
0	0	.25	0	0	.02	0	0	.25
0	1	.25	0	1	.48	0	1	.25
1	0	.25	1	0	.48	1	0	.25
1	1	.25	1	1	.02	1	1	.25

Figure 2

In this situation we expect that decompositions which do not decouple or separate variables v_2 and v_3 can provide useful information regarding the structure of the overall relation as represented in the data. To deal with such situations it is thus necessary to consider behavioral representations of the second type as described earlier in this section, those which augment the samples which occur by their probabilities of occurrence.

In the second behavioral description the determination of meaningful structure systems is extended by the included consideration of the probability distribution, though the fundamental question remains: to what extent is it possible to consider the overall system of n basic variables as consisting of subsystems, each of which involves fewer than n of the basic variables?

It would be possible to extend the cylindrance example just described to include the consideration of probabilities. In such cases we extend the definition of behavior – both of the overall system and of subsystems – to include probability distributions associated with the sets of samples. For the overall system the probability distribution is determined as above in (2). For the behavior of b_i, associated with subsystem $e_i \in P(V)$, we again use the projection as in (8) but augment each sample with a weight determined from those given in (2). That is, we associate with each $x \in b_i$ the weight $p(x) = \Sigma p(d_j)$ where the sum is taken over all d_j such that $\mathbf{Pr}_i\{d_j\} = \{x\}$.

To determine the acceptability of a particular E_c, we again take:

$$\bigcap_{b_i \in E_c} \mathbf{Sp}(b_i).$$

Call this set T. It is now necessary, however, to determine for each $t \in T$ an associated weight, $h(t)$, and compare this with the weight of that sample originally assigned as in (2). This is accomplished by multiplying $p(x)$, as determined in (2), by $p(\mathbf{Pr\text{-}}_i(\{t\}))$ where $\mathbf{Pr\text{-}}_i$ indicates the projection onto all those coordinates which are not elements of e_i. In the general case it will be necessary to calculate $\# E_c$ weights for each element of T. The $\# E_c$ sets of weights so calculated could be checked against the original weights and a match for any complete set could be said to justify consideration of the overall system as one whose modified cylindrance was c.

Although this procedure generates a modified cylindrance of 2 for the example in Figure 1, this property is not general and almost any slight modification of the probabilities negates the result. It is possible to formulate a somewhat complicated measure relating $h(t)$ and $p(t)$, but at this degree of complexity almost all intuitive and interpretive connection to the original is dissipated.

While it is useful to consider the cylindrance for non-probabilistic behaviors it is more fruitful to consider an alternate approach to specification of lower-order relations, and alternate structure evaluations specifically related to meaningful investigative purposes for probabilistic behaviors.

In this approach we relax the requirement that each subsystem involve the same number of variables and also allow that any number of variables of a given subsystem may be shared by another. Following Klir and Valach (1965) and Klir (1969), we call such shared variables *coupling variables.*

We thus again consider a structure system to be a set of subsystems with associated behaviors, where each subsystem is represented by an element of the power set of V, that is, by a subset of the basic variables. Since every variable must be included in some subsystem and since we do not consider directions of the variables, we can consider each potential structure system to define a compatibility relation on V, that is, a reflexive and symmetric binary relation. For example, a system investigation may involve the five basic variables $v_1 \ldots v_5$. This may be diagrammatically represented as follows:

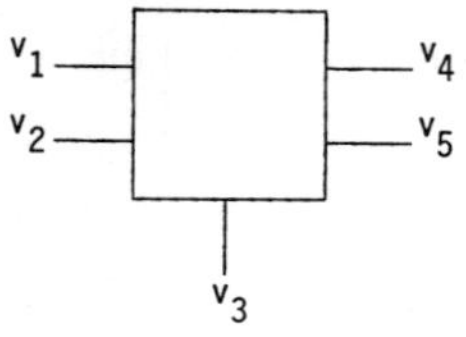

Figure 3

The potential decomposition (refinement or structural disintegration), giving a structure system which involves the subsystems $\{v_1, v_3\}$, $\{v_3, v_4, v_5\}$, $\{v_2\}$ may be pictured as in Figure 4.

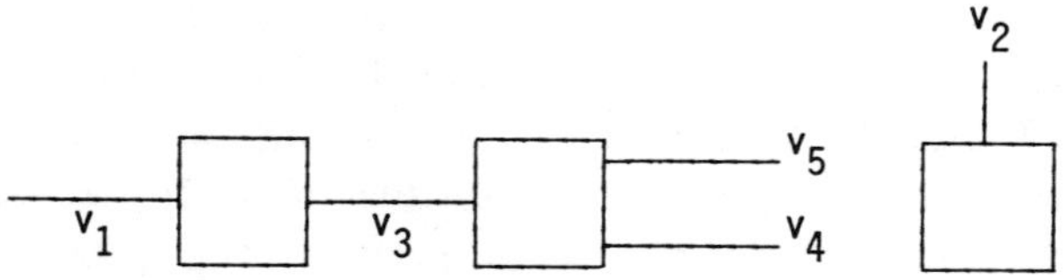

Figure 4

For many characterizations and problems associated with the determination and evaluation of structure systems it is useful to utilize the binary relation representation of the structure system, where two variables v_i and v_j are related if and only if they belong to at least one of the subsets of v through which the structure system is defined. With this representation it is very simple to test for certain conditions or criteria which will be relevant to the evaluation of different structure systems from various perspectives.

For example, in the attempt to determine a structural model of a system it is often desirable to require a crisp decomposition in the sense that each subsystem must directly relate some variables which are not related through any sequence of subsystems (or, which are not otherwise related directly *or* indirectly). Thus with this restriction each subsystem of a potential decomposition must provide *some* unique information. In terms of the binary relation which is associated with a possible structure system this translates simply into the requirement that the transitive closure of the relation relates at least two variables which are not related in the transitive closure of the relation obtained by eliminating from the relation the ordered pairs represented by any subsystem (or element of the structure system).[3]

Given the various decompositions which may be made of an overall system, and which satisfy the above criteria, the simplification task is to choose one which simplifies the system as much as possible and which still retains chosen characteristics of the system within tolerable limits. That is, each decomposition, through removal of certain interactions among variables, potentially discards important information about the overall system. As Klir has observed: 'The question is whether or not the whole structure system represents correctly the data system' (Klir, 1976). Depending on the purposes of the investigation, which will be translated into our framework as objectives and constraints the phrase 'represents correctly' may have different interpretations and signify different characteristics of the system. The function of a general systems framework is to elaborate various issues involved in different interpretations and to develop mechanisms by which these various interpretations can be operationalized.

Klir (1977) has accomplished this operationalization at a basic level, that in which 'correct representation' is interpreted as the direct ability to reconstruct statistical structure of the overall or total relation from the decomposition which structure represents. In this case a distance is defined which measures the difference between the overall relation (with probability distributions) which is implied by the (hypothetical) structure system and the corresponding empirical relation based directly on the data system. (This is illustrated in more detail in the following chapter.) The procedures developed for implementing this decomposition (see Klir and Uyttenhove, 1976) give explicit recognition to the fact that from the general system perspective it will often occur that no structure – that is, no non-trivial decomposition – will be able to capture the overall system exactly. The value of such procedures, however, is that they present the investigator with the choice and determination of the relative tradeoff between simplification and discarded or lost information and do not explicitly make this choice outside of context through assumptions which are embedded in a technique. This is an important methodological issue and again points to the fundamental scientific and general systems objective to remain as close as possible to the given system characteristics and adapt

3. For further elaboration of the utility of the relation representation, especially with respect to development of computer implemented approaches to structure modelling, see Cavallo and Klir (1978b).

tools and procedures to fit these situations rather than to approach the investigation with a predetermined model or form and adapt the object to it.

We also observe that the subsystems derived by general systems decomposition techniques as described here serve the same fundamental goal as traditional techniques such as factor analysis. Indeed, the subsystems associated with each structure system are analogous to factors derived from factor analysis and represent all the potential theoretical benefits derivable from parsimonious description (this connection is also further elaborated in the next chapter). They are not, however, based solely on the (generally inadequate) interaction among only pairs of variables. In this regard it is easy to construct illustrative examples of simple systems with constraint for which no constraint (or interaction) is exhibited between any pair of variables. Thus, techniques based only on pairwise interaction (measured by correlation coefficients) would necessarily detect no structure (see Ashby, 1965a; Krippendorf, 1978).

5.4. Separability in Design

The considerations of decomposition which we are encompassing in our definition of structure systems are fundamental to the activity of modelling and, more significantly, that of modelling with a purpose. That is, in defining a system on an object of investigation, the purposes of the investigation often involve the potential for action by the investigator (that is, for choice from among a set of alternative actions) with the intention of affecting the status of certain attributes which may be considered as outcome or goal specifications. To formally describe this overall process we would embed the considerations as given here in a more encompassing description involving an extension of the defined system by the set of potential actions. For our purpose here, however, we merely refer to the interpretation (see Ashby, 1967; Conant and Ashby, 1970) and observe that the task of coordination of choices with system behavior can be significantly simplified if the system can indeed be meaningfully considered as being composed of simpler parts.

With this emphasis an alternative interpretation to that given by Klir, in the determination of the validity of different decompositions, suggests itself. Situations of this character are called dynamic design systems by Churchman (1971) who discusses the possibility of a designer focusing emphasis on parts rather than on the whole system. Churchman argues that the 'crucial point in the design is whether one can recognize the unsatisfactory state of a part without having to study the entire system in depth.' If we call 'the principle by which a part is changed a "transformation function"' we observe that this is 'equivalent to asking whether the transformation functions [of proposed subsystems or parts] are functions of the prior states of [the overall system], or merely functions of a subclass of the parts . . . The strongest form of the separability principle . . . states that the tranformation functions of a part are functions of the prior states of the part only' (Churchman, 1971, pp. 64-65).

Churchman discusses the issues quite clearly and reinforces our general conception that acceptable decompositions can transform essentially impossibly complex situations into ones which are easily manageable. It is also important to observe that this orientation recapitulates the scientific or analytic emphasis on the need to in fact study parts. This reinforces the comments of Weinberg and Ashby referred to in Chapter 2 which recognize that information processing constraints do impose necessary restrictions on what can be accomplished by an unrestrained 'wholistic' approach. This is only to restate that the analysis/synthesis contradiction is not one which can have a general ultimate resolution which favors an emphasis solely on one aspect. The important ingredient which general systems research has served to emphasize regarding analytic aspects is that care should be taken – and that this must consciously be integrated into the investigative framework – to include consideration of the 'whole' in the determination of what will constitute the 'parts' chosen for study.

In implementing this consideration with respect to the description given by Churchman, however, we recognize – as we have described in the section on generative systems – that questions which depend on transformation functions – that is, on the dynamics of the system or its parts – are not easily resolved. For systems which are of most interest to social scientists it is often not possible to determine the transformation functions, or at least not meaningful to give their specification. We describe here an approach which captures the essence of Churchman's concerns and is realistically applicable to observationally defined systems but does not require complete knowledge of the actual 'transformation functions.'

In this derivation we assume that at any moment of action the designer or policy-maker can know the current state of the overall system (that is, the current appearances of all attributes). With respect to potential actions or decisions the agent is interested in the extent to which these actions may be taken in consideration of parts of the system rather than of the overall system – that is, of the full set of basic variables. If we denote the current state of the overall system by S then, even presuming that the investigator considers the overall relation, there will generally remain an inability to deterministically predict the next state. For convenience we consider S' to include only current values of the basic variables. There is no theoretical necessity for this restriction but data considerations often demand it.

If we denote the 'next' state from any reference parameter value by S' then the conditional entropy:

$$H(S'|S) = -\Sigma\Sigma p(S,S') \log p\ (S'|S), \tag{10}$$

where S and S' are taken over the whole of the process under investigation, is a measure of the uncertainty which the investigator must face even if taking account of the full spectrum of object-defining attributes. Assuming that the overall system is retained as the focus of interest, but that the totality of attributes represents an information processing overload on the part of the designing agent,[4] then separability into parts is an inescapable necessity.

Here the primary difference between what is regarded as the systems approach and the traditional analytic paradigm is fairly clear, emphasizing that the choice of parts which are to be given separate attention cannot be made in an arbitrary or intuitive manner, choosing as elements those which a particular orientation or reference frame may have accustomed the designer to accept as the 'natural elements' (Rosen, 1977). That is, the choice of parts should only be made after fundamental consideration of the overall system, as well as consideration of the purposes of the investigation. The fundamental contributions of Ashby's cylindrance, Atkin's Q-analysis, and Klir's identification of generative structures are the provision of operatory schema within the context of generally meaningful representation schemes, which embody this systems emphasis by incorporating information which may be present only in the overall relation.

In extending this to the context of the interpretation under consideration, we utilize the lattice of structure candidates described by Klir (1976) and Klir and Uyttenhove (1976) and limit our attention to the class of structures

4. The relevance of some sort of information-processing limit has been discussed from various perspectives. See, for example: Miller, 1967; Ashby, 1965b; Bremermann, 1962; Simon, 1962.

dealt with in that work (see also Cavallo and Klir, 1978b). For a system defined by three basic variables, v_1, v_2, and v_3, the possible structures are as shown in Figure 5.

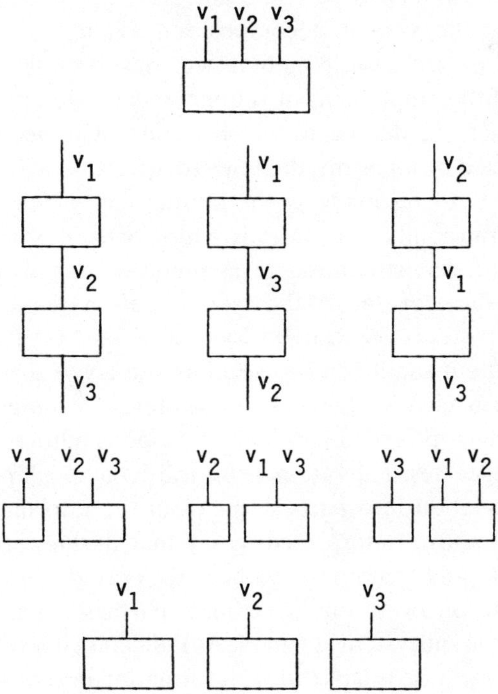

Figure 5

In this figure the individual blocks are taken to represent subsystems and the labelled lines connected to each block represent the variables associated with that subsystem. With three variables, of course, the options for subsystem formation are few and Figure 5 in fact lists each of the seven possibilities – including those decompositions which involve coupling variables connecting two subsystems. In the more interesting cases utilizing larger numbers of system-defining observables, the number of possible refinements (or structure systems) rapidly increases to thousands or millions, and the possible forms of the available structure systems become much more interesting (examples for a larger number of variables are included in Chapter 6). An obvious feature of the decompositions is that large numbers of different decompositions are different only in the sense of a permutation of the variables. For example, all of the structure systems in each line of Figure 5 are of the same 'structure type,' and this fact, along with the lattice structure which the structure systems possess, makes procedures dealing with the full complement of structure candidates retain computational feasibility (cf. Cavallo and Klir, 1978b).

To exemplify the system decomposition evaluation and acceptance criterion which we are describing here we consider only a three-variable system and its full complement of refinements as pictured in Figure 5.

In determining the degree to which a particular decomposition may acceptably be used by an agent, the relevant quantity or criterion by which the judgement is to be made is the amount by which the uncertainty associated with the ability to act in anticipation of the next state – if the parts and couplings of that structure are assumed – exceeds the uncertainty associated with the next state of the overall relation (depicted at the top of Figure 5). This difference can be looked at as the overall next-state information content sacrificed by assuming the given simplification.

The basic approach described here is relevant, for example, in attempting to determine the optimal allocation of responsibilities among various committees or agencies. The rationale for the decomposition or refinement of the overall relation into parts stems from the information processing limits described above, interpreted to say that during a period of action represented by a single parameter value (time period) a given agency need only concentrate on occurrences within a particular part of the system represented by the subsystem assigned. As indicated, it is often meaningful, however, to incorporate full past state information into this determination. In this case such information could refer to reports received on the outcomes of all the defining attributes for the previous period. As we show, the assumption of the mutual availability of such reports determines different optimal structures than is the case if only past information regarding the given subsystems is considered.

We denote the set of subsystems or parts associated with a given decomposition, **d**, by $E_d = \{e_i\}$ where each e_i is again a subset of V, where $\cup e_i = V$, but where the further criterion that variables not be related both directly and indirectly is satisfied.

Since we allow that past information of each subsystem be available to each agency, we are interested in the quantity $H(S'_i|S)$ for each subsystem, where S'_i representes the next-state of subsystem e_i, and in the quantity $H_d(S'|S)$ which denotes the total uncertainty in the next state of the system, assuming the decomposition **d**. In cases where the parts are not coupled, as for example in levels 2 and 3 in Figure 5, $H_d(S'|S)$ is merely the sum of the

$H(S'_i|S)$. When, as in level 1, however, parts share coupling variables, then we assume that the information associated with determination of the next state of these variables is available to all subsystems so coupled. Thus the calculation of $H(S'_i|S)$ must be made in conjunction with the next-state entropy of those elements with which it is coupled (see also Broekstra, 1978). This involves a single calculation of the next-state entropy for the coupling variables, and allows the utilization in the remaining coupled subsystems of the values of these variables as conditioning information.

Let two subsets of V, e_i and e_j, represent two coupled subsystems. Then $(e_i - e_j) \cup (e_j - e_i)$ is the set of variables which are not directly related and $e_i \cap e_j$ is the set of coupling variables shared by the two elements. Let $H(S'_{ij}|S)$ represent the total next-state entropy for these two elements, and denote by S'_{j-i} the next state of those variables included in subsystem e_j but not in e_i, and by $S'_{i\cap j}$ the next state of the shared variables. Then:

$$\begin{aligned} H(S'_{ij}|S) &= H(S'_i|S) + H(S'_{j-i}|S, S'_{i\cap j}) \\ &= H(S'_j|S) + H(S'_{i-j}|S, S'_{i\cap j}) \end{aligned} \quad (11)$$

For computational simplicity in calculating the next-state entropies associated with each decomposition we make use of the identity:

$$H(Y|X) = H(X, Y) - H(X) \quad (12)$$

to write:

$$H(S'_{ij}|S) = H(S'_i|S) + H(S'_j|S) - H(S'_{i\cap j}|S) \quad (13)$$

Thus, calculation of the next-state excess entropy associated with each decomposition – that is, $H_d(S'|S) - H(S'|S)$ – can be effected by additions and subtractions among initially calculated next-state entropies associated with subsets of the set of basic variables.

To illustrate the procedure, and for comparison with other possible criteria for evaluating decompositions, simplifications, or 'factor' determinations, consider a system defined by three binary variables for which there is defined a totally ordered parameter space. Let the record of observations, or data-gathering function f, be given by the data array in Figure 6, where each column represents a not necessarily distinct element of Imf and where the column identifiers are understood to be the elements of the parameter space.

```
[11000000110100000000110000000010101010000111111111110
 10100001101111100001010111100100011110110111111111
 10110101001000000101011000000111010000100111111110

 11011101110111111110001001000100111111110010011110
 00101011001011111111101111010010111111101000110100
 10111010101111111100000000101101111111000111100011

 0010000001010011111000111010000001011001100000000001111]
 0001001001000111001001111001000001001000100001111111
 0001101101110000101101110111001011011101100100001111]
```

Figure 6

For simplicity, the current state of the system is understood to involve only current values of basic variables. Table 1 lists the relevant next-state entropies where $H(v_1' \, v_2' \, v_3' \mid v_1 \, v_2 \, v_3)$, for example, indicates the average amount of uncertainty which can be expected, based on past observations, regarding the next-state of the system considered as a whole (that is, taking into account all interactions among the variables or, alternatively, taking full advantage of existing redundancy) given the current values of all variables. Similarly $H(v_1' \, v_2' \mid v_1 \, v_2 \, v_3)$ represents the uncertainty regarding only the next-state of v_1 and v_2, conditioned by knowledge of all current values.

Table 1

$H(v_1' \, v_2' \, v_3' \mid v_1 \, v_2 \, v_3)$	= 1.690
$H(v_1' \, v_2' \mid v_1 \, v_2 \, v_3)$	= 1.391
$H(v_1' \, v_3' \mid v_1 \, v_2 \, v_3)$	= 1.332
$H(v_2' \, v_3' \mid v_1 \, v_2 \, v_3)$	= 1.295
$H(v_1' \mid v_1 \, v_2 \, v_3)$	= .783
$H(v_2' \mid v_1 \, v_2 \, v_3)$	= .721
$H(v_3' \mid v_1 \, v_2 \, v_3)$	= .797

To illustrate the effect of consideration of information regarding all variables of the overall system as conditioning information with respect to the expectations for the separate subsystems of a given structure, we also evaluate structure candidates without making this assumption. That is, we consider that only information regarding values of the variables represented in a subsystem is used in determination of the next-state entropy of that subsystem (this is more in keeping with the description given by Churchman). The relevant figures are given in Table 2.

Table 2

$H(v_1' \, v_2' \mid v_1 \, v_2)$	= 1.739
$H(v_1' \, v_3' \mid v_1 \, v_3)$	= 1.787
$H(v_2' \, v_3' \mid v_2 \, v_3)$	= 1.672
$H(v_1' \mid v_1)$	= .953
$H(v_2' \mid v_2)$	= .983
$H(v_3' \mid v_3)$	= .990

Using the figures of Table 1, we are able to calculate the quantity $H_d(S'|S)$, for each structure candidate of Figure 5. These results, along with other calculations, are presented in Figure 7.

On the left of Figure 7 is reproduced the diagrammatic representation of structure candidates of the form we are considering for a three element system. The figures in column 1 give the quantity $H_d(S'|S)$, as described, for each structure. The figures in column 2 give the calculated next-state entropies for each structure candidate, but using the figures of Table 2. The final column gives the distance measure for each structure candidate as described in Klir (1976)[5] and using a complete mask of depth 2, thus using the same data at each measuring point as with the other two criteria. The circled numbers in each column for levels 1 and 2 represent the ordering of the structure candidates at that level according to the criterion represented in the column.

Although the example was arbitrarily generated, certain important differences appear which clearly illustrate the significance – for the generation and solution of systems problems – of purposes, requirements, and constraints of the context of the investigation. At level 1, for example, the structure candidate which would create two agencies or responsibility areas coupled by variable 2 is the worst using the next-state entropy

5. The distances of column 3 were calculated using a set of programs developed by Klir and Uyttenhove (1976).

Level				
0)	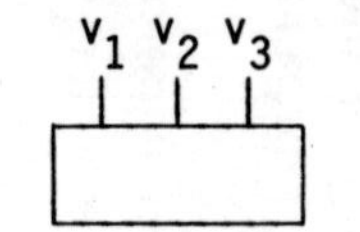	1.690	1.690	0
1)	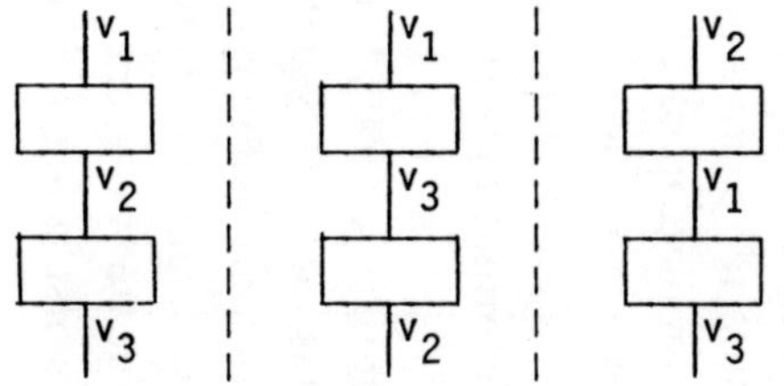	1.965 ③ 1.830 ① 1.939 ②	2.510 ① 2.607 ② 2.646 ③	.7746 ② .7662 ① .8779 ③
2)	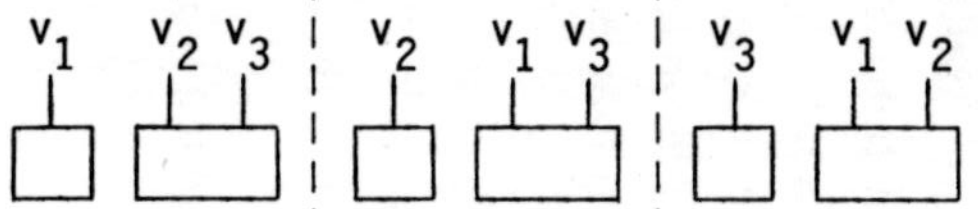	2.078 ② 2.053 ① 2.188 ③	2.625 ① 2.770 ③ 2.729 ②	.9050 ① .9940 ③ .961 ②
3)	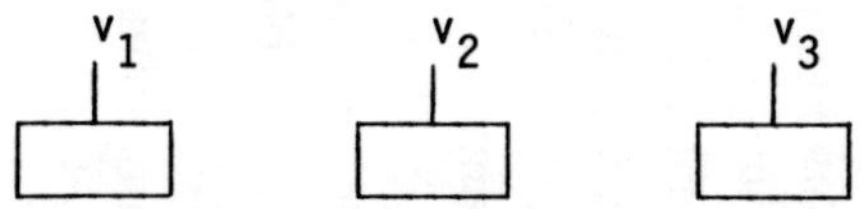	2.301	2.926	1.0289

Figure 7

criterion and assuming access and consideration of end-of-period reports from all agencies. Without such access, however, this structure candidate emerges as the best in consideration of next-state entropy (column 2) and as second best using the distance-measure based on sub-masks (column 3).

At level 2, the evaluation of candidates using the uncertainty criterion without communication of past-state information among subsystems is order-isomorphic to that using the distance measure. In each of these cases, however, the candidate evaluated as worst (that which isolates variable 2) is in fact the best when past information may be exchanged. This example, then, clearly illustrates the need for ampliative consideration of the 'sense' that given criteria have with respect to the purposes for which a systems investigation is being undertaken.

The preceding considerations emphasize again certain fundamental features which are critical to an operational general systems methodological framework. These are:

1. The core concepts – for example, behavior and structure – must be defined in a general enough manner to encompass wide varieties of specific utilizations of them.
2. Any given utilization – that is, any given overall problem situation – will invariably involve observations or pseudo-observations (a priori specification of behavior as in design situations). Mathematical and computational expressions and procedures represent a most efficient means for handling these observations. The framework must thus allow for a classification of problem situations which associates proper operatory schemata – generally utilizing results of research in abstracted level languages – with the situations, and which properly takes into account the purposes, objectives and constraints of the investigation.
3. Operatory schemes are themselves integrally a part of the epistemological process, but must not be separated from the overall epistemological level concepts among which they provide the means for knowledge growth.

5.5. Summary of Fundamental Concepts

In the last three chapters we have developed the argument that the fundamental concepts associated with Klir's epistemological-level hierarchy of systems constitute a basis for a comprehensive framework which can serve as a methodological language for general systems research. A summary of the system levels is presented in Figure 8.

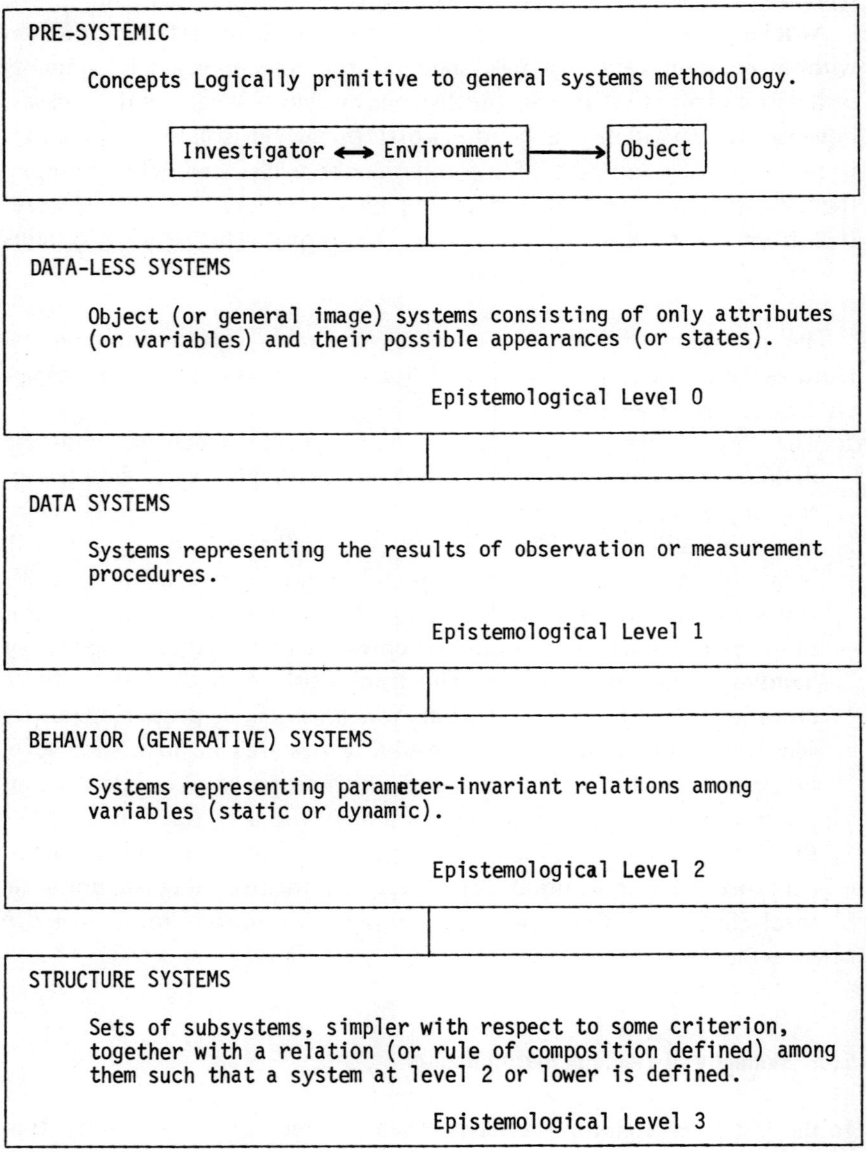

Figure 8

We also mention that it is possible to extend this basis by considering systems with varying structure where the structured variations are considered to relate to variations in the parameter space. Such extensions have been called metasystems (Klir, 1978c) and procedures related to their investigation are currently under development (e.g. Uyttenhove, 1978).

A major emphasis in the elaboration we have presented in the preceding chapters has been the illustration that operational procedures which constitute an important core of social scientific research relate to the system levels in a natural way and, further, that the framework itself motivates the development of procedures to answer questions which are important for many applications.

We have indicated that the utility of the framework for scientific research depends on a fundamental integration of these procedures, as well as of objectives and constraints relevant to given research situations. The next chapter provides a development of the framework which accomplishes this.

6. GSPS

Indeed from our present standpoint, physics is to be regarded not so much as the study of something a priori given, but rather as the development of methods for ordering and surveying human experience.

Neils Bohr
Atomic Physics and Human Knowledge

One thing is sure, though, that the methods themselves must change if aims change.

Henri Cartan
Nicolas Bourbaki und die Heutige Mathematik

The last three chapters have presented fundamental concepts which are representative of investigative processes and have described the manner in which these concepts lead to a natural classification of general systems. Throughout this development a major emphasis has been placed on process-oriented aspects of knowledge acquisition, an emphasis which demands an augmentation of static conceptions to include relations and movements among the systems levels which the concepts motivate. This is achieved in this chapter through the development of a conceptual framework referred to as general systems problem solving.

In keeping with the primary focus of this book this material is presented in Part 1 concurrently with considerations of special importance to the investigation of social systems. Its applicability and utility is illustrated in Part 2 through an example drawn from an important social science research area, that of the study of intranational (or domestic) conflict.

Part 1: An Organized Methodological Framework

6.1. Knowledge as Process

In relating this framework to research in the social sciences it is necessary to contrast our position, which places the primary emphasis on *activity* with positions which place the major stress on *theory* and which expend considerable energy on the formulation of conditions which must be fulfilled if one is to have a 'valid theory.' In this regard this framework grants secondary status to the 'finished research report' aspects (Hanson, 1958, 1965) which are often given a major emphasis in the writings of social scientists (e.g., Meehan, 1965; Holt and Richardson, 1970). For example, rather than argue relative merits of *falsification* versus *verification*, we prefer to recognize with James (1907) that the process aspects which are implied in the *-fication* part of the words relegate controversy over the prefixes to mainly scholastic status.

While we agree that non-research oriented explications of what 'scientific theory' must consist of can serve a useful critical function, certain recurring themes in social science writings – for example, wistful accounts of why sociology has not yet 'achieved scientific status' (Lachenmeyer, 1971) – imply an unduly negative perspective. The orientation of this framework is much more in consonance with the judgement of Przeworski and Teune that the stress in many writings on what Melville (1976) refers to as a 'high-Carnapian language game' grossly underestimates the importance of actual research and that 'if such philosophy of science books are understood and taken literally, research is likely to be paralyzed' (Przeworski and Teune, 1970, p. x).

The static accounts of theory and science – often based on formalist or positivist accounts of models of physical science – are suspect on three criteria:

First is the previously discussed perspective that social and humanistic systems do indeed represent phenomena of a different order of complexity than those of the physical sciences (Hayek, 1964; Bohr, 1963; von Neumann, 1951).

Second, placement of a major emphasis on the hypothetico-deductive method is based on traditions which are currently recognized as deficient in light of their assumed ability to disregard such 'intangibles' as creativity, insight and use (Wittgenstein, 1956; Castonguay, 1972; Heisenberg, 1958). Even within mathematics, for example, Thom argues that 'there is no rigorous definition of rigor... a proof is rigorous if, in any sufficiently instructed and prepared reader, it creates a state of evidence which imposes conviction' (in Castonguay, 1972), and Dieudonnè eschews the 'heap of axiomatic trash dumped every year by would-be mathematicians on the unhappy public' (Dieudonnè, 1964).

The third and probably most important criterion tending to lessen the value of static theory critiques is that the accounts they are based on are rarely accounts given by working researchers. The latter are much more willing to admit of the limits to formalization, the importance of natural language, and the recognition, for example, that 'in the development of natural science [it] will be rather the exception than the rule... [that a] sentence belongs to a closed system of concepts and axioms' (Heisenberg, 1958, p. 85).

6.2. 'Theory' and Data in the Social Sciences

A related theme which often appears in literature which emphasizes the 'hypothetico-deductive' method is an almost denigrative attitude toward the actual utilization of data. In recent decades most of the social sciences have undergone a significant boom in quantification and the use of quantitative methods and approaches (e.g., see Jones and Singer, 1972; Deutsch, 1973; Heise, 1975). Some of the criticisms directed toward the use of these approaches are constructive and justified, especially given the possibly overenthusiastic and often arbitrary uses of methods which seem to have been undertaken merely because these methods exist and without attention having been given to limitations which methods possess (e.g., see Einhorn, 1972; Forbes and Tufte, 1968).

But these criticisms often extend unreasonably to question the very compilation of data and systematic recording of observations. These critical extensions often give a decided impression of fishing-expedition critique for critique's sake. Job and Ostrom (1976), for example, criticize Singer's 'a-theoretic' compilation of data in connection with the Correlates of War project at the University of Michigan. The a-theoretic charge stems from the fact that a well specified theory was not in evidence before the actual collection of the data. Singer (1976) quite adequately responds that he has in fact 'been theorizing about the causes of war for about three decades' (p.

129) and that it was the very lack of an acceptable theory which led to his decision to emphasize the collection of data which is as reliable as possible.

The situation here shows a clear similarity to that in the field of elementary particle physics where Heisenberg (1976) has compared the yearly increase in data on masses and quantum numbers of particles to surveys in astronomy. Though Heisenberg naturally would desire a dynamical theory, he nevertheless recognizes that 'new experimental results are always valuable, even if they only enlarge the data table.'

Regarding this conflict between theoretical and empirical concerns in the social sciences, statements often appear to be arguing that investigations must be based on either one or another of two antithetical approaches, either that:

1. data 'generates' theory, which in its extreme does give the impression of a belief in the possibility of a 'theory machine' which need only be fed data to crank out theories; or,
2. theory can exist independently of data, which in its extreme implies the ability to constrct 'theory' without ever having observed what the theory purports to describe.

A major point of this book has been to argue that neither of these two extremes can constitute an acceptable epistemological basis for scientific inquiry, that the potential for knowledge growth resides in the tension between theoretical and empirical concerns, and that this tension is primarily represented or captured in the concern with methodology. It is precisely this recognition that underlies and gives major significance to the work of Zadeh, Atkin, Wymore and others as described in Chapter 2 and which has motivated the development of the framework which is described here.

This recognition has of course also been made from within the social sciences, but a major aspect of our argument has been that the scope of the issues involved warrants the consideration of many of these issues in a more general and comprehensive manner than can be achieved under the constraint of primary concern with specific context. Equally, we emphasize that this consideration must be made in a manner which is still organically connected to general aspects of various contexts.

Evidence of the need for a general and comprehansive methodological framework in the social sciences is given, for example, by 'faddish attachments' which tend to develop around particular research tools or techniques which happened to have found a particularly innovative use (see

Graham, 1971; La Palombara, 1970). The difficulty, of course, is not with the use of any technique but with the recurring tendency to shift the emphasis from the natural concerns of the social sciences to the technique being used, as with Goethe's perception reported at the end of Chapter 4 regarding the use of mathematical formulae.

As we have developed the conception of general systems research the main point of view has been that it should be considered as a language or communication channel between specific research and abstracted techniques or constructs, one which is particularly suited to serve as the mediating function between theoretical and empirical description.

The next section provides a means by which the danger of methodological biases and faddish attachments can be minimized. This is accomplished by defining a mechanism which allows a spectrum of research orientations, which in fact specifies a 'catholic perspective' which may be used for social science research (Kaplan, 1964) but which involves a necessary codification of this catholicity, thus allowing for systematic rather than haphazard or random utilization of the major elements which are involved.

6.3. General Systems Problem Solving

6.3.1. GSPS

The dynamic and constructive aspect of the epistemological process, emphasis of which constitutes the main element of this framework, suggests a characterization of investigation and discovery which is similar to that given by Pask and which recognizes that 'a human being does not so much respond to stimuli as interpret certain states of his environment as posing *problems* which he makes an attempt to *solve*' (Pask, 1969). Pask's comment is in keeping with the results, which we have reported, of Piaget's investigations into the workings of the growth of knowledge and is offered as an alternative to the naive behaviorist view of man merely as one who 'reacts to stimuli'.

Similarly, Tukey utilizes an emphasis of problems as an alternative to the equally naive view, extreme from the opposite direction: 'To judge from published books and articles, experimental statistics has grown by finding tools somehow, and then running around using them... Why has experimental statistics not been more obviously concerned with problems?' (Tukey, 1954). In considering this situation Tukey emphasizes the importance of difficulties in the *identification* of problems relative to that of their *solution*, and argues that 'it is appropriate to be as systematic as we can about... problems'.

In the manner which we conceive the need for both a conceptual scheme and operatory schemata for movement among the concepts involved in the scheme, the notions of problem and problem solution are well suited to encompass these aspects. We thus effect the integration of these notions into the overall framework through the development of the notion of a *general systems problem solver* (GSPS).

The main intention of GSPS is to characterize fundamental methodological (rather than theoretical) situations as *systems problems* and to do this in a manner which is suitable for use by different social and behavioral sciences. It is thus necessary to incorporate consideration of purposes of the overall investigation – of which the systems problems are a part – in a manner which involves a meaningful association of methods or methodological tools to the problems so defined. Since the framework is intended as a linguistic and operational aid it is important to allow for as precise a specification as possible of the concepts involved. (Cavallo and Klir, 1978a, introduces some of the material of this section and also contains a complete specification of an implementable software package using small subsets of the possible system types, requirement types and procedures. Zeigler, 1978, discusses some issues from a slightly different perspective relating to the definition and implementation of an extension of modelling capabilities through automated procedures. These approaches which organize or automate and integrate certain fundamental and formal aspects of the investigative process should be clearly distinguished from attempts to reproduce or ostensibly remove the need for the knowing subject.)

6.3.2. SYSTEM TYPES

The first concepts are based on the characterization of systems and the notion of *system type*. We thus recognize that each level in the hierarchy requires certain well defined information for communicable classification, as has been defined in Chapters 3 to 5. To encompass diverse situations we also allow that each basic system type so defined may be modified by the presence of certain other information which is relevant to system classification and which may arise in specific investigations. This information may be, for example, specification of an initial state, classification of variables into input and output, or part of the information necessary to categorize the system as one at a higher level.

Similarly, provision must be made for the inclusion of criteria which encompass various methodological distinctions which are made in different research situations. These generally refer either to the variables or to relations which are involved in the system specification. Some examples of these distinctions which commonly arise are:

Classification by Variables
- nominal/ordinal/metric
- discrete/continuous
- well-defined/fuzzy
- unordered/partially ordered/totally ordered parameters

Classification by Relations
- deterministic/probabilistic
- well-defined/fuzzy
- general/structural (linear, etc.)

System types are thus defined by specification from among certain criteria which express the totality of recognizable types, and which criteria may be considered either as epistemological (such as the choice of level) or as methodological (such as a statement as to whether the variables are nominal or ordinal scale). The examples given are, of course, not inclusive and may be augmented by other criteria which are deemed acceptable.

6.3.3. PROBLEM KINDS

As we have emphasized, and as illustrated in Figure 9, general systems research – and more specifically GSPS – should be considered as a communication channel between specific research and abstracted techniques or conceptions which are not tied to semantic reference. As such GSPS is intended to provide a means for organizing aspects of overall research situations in a manner consistent with a broad spectrum of considerations from the abstracted languages.

The major organizational feature of GSPS which achieves this linkage is the expression as systems problems of various activities oriented toward the investigation of system aspects of a situation. The epistemological-level based system type as described forms the core of the specification of a system problem. We will define systems problems in terms of these levels in one of two ways:

1. As a transition from a system identified at one level to another expres-

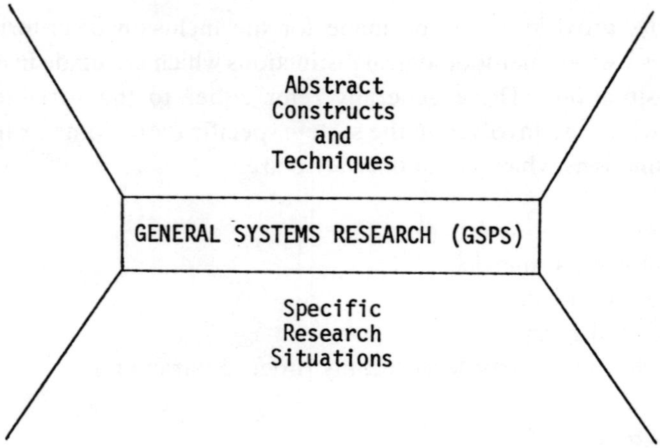

Figure 9

sion, possibly at the same level, but more importantly to one at a lower or higher level (e.g., given a behavior system, determine a structure system where each subsystem involves no more than three variables and which minimizes the next-state entropy; given a system defined by its state-transition matrix, determine a homomorphic simplification involving no more than eight states).

2. As an expression of the relation between two given systems (given two behavior systems, determine which exhibits the most redundancy or constraint among the variables or, equivalently, with respect to a certain decomposition criterion determine which system exhibits the most structure).

We refer to problems which are defined to account for these two different situations as *problems of the first kind* and *problems of the second kind*, respectively. Complete specification of a particular problem of either kind requires certain refinements which we introduce next, but it is important to emphasize that the resulting classification of problems is only a preliminary objective. The major role which the classification serves is to provide a means for either: (a) the association of existing abstract conceptions from, for example, mathematics and computer science to the problems defined which will serve as methodological tools for problem solution; or (b) the detection and highlighting of those systems problems for which adequate solution procedures do not exist, thus providing the motivation for abstract research which will in fact be of immediate use.

As will become clear through the complete definition of systems problems, *problem kind* represents only a part of our conception of problem. The augmentation of objectives and constraints specific to particular investigations will obviously have a major effect on the suitability of given methodological procedures. But, even more important, our conception of problems, as described in Chapter II, leads to the recognition that even for given completely specified systems problems, there need not be – in general – a single procedure which 'solves' that problem. Different procedures often imply different reference frames which can give different solutions and different information; these imply enrichment of knowledge with respect to the object through the use of complementary perspectives which are, nevertheless, methodologically integrated through use of the framework.

6.3.4. REQUIREMENT TYPES AND PROBLEM TYPES; PARTICULAR SYSTEMS, REQUIREMENTS, AND PROBLEMS

In most investigative situations which generate particular systems problems we can expect that the systems problems so generated include objectives or constraints, the consideration of which constitutes an essential part of the problem. For example: the common social science practice of 'dimension' determination through the use of n-dimensional P-mode factor analysis can be considered as a problem involving transition from a behavior system (specified in the generally unsatisfactory manner of correlation coefficients between pairs of variables) to a structure system (the set of dimensions) where the number of subsystems is specified beforehand; in attempting to derive a difference equation model to describe time series data (transition from data to behavior system) the context of the problem may demand that the difference equations be no more than first order; in determining an optimum rule for data transformation (transition from source system to source system, or data system to data system) the data system must represent a linear function from the parameter space to the basic variable space.

The set of all well-defined objectives and constaints which GSPS is able to recognize is called the set of *requirement types.*

Given the system types and requirement types, then definition or characterization of a *problem type* involves the listing of two system types – distinguishing one as *initial* and one as *terminal* – and a subset of the set of requirement types (thus a given problem may involve more than one objective or constraint).

While system, requirement, and problem types represent the fundamental categories which are relevant to GSPS, we distinguish between the types and particular embodiments of them. The latter are called *particular systems, particular requirements* and *particular problems.* Thus, for example, the statement:

compare two behavior systems, with basic variables partitioned into dependent and independent variables, to determine which involves least uncertainty in the dependent variables conditioned by knowledge of the independent variables

identifies a given *problem type*, while a description of the behavior systems by specification of the relations which define them would constitute a *particular problem of the second kind.*

In the following section we provide a formal description which ties together the concepts relevant to GSPS. The goal regarding the development of GSPS is to explicitly define and extend the relation between the set of problem types which are *identifiable* (that is, which make sense) and the set of methodological tools or procedures which can be applied to their solution. In this regard we distinguish those identifiable problem types which participate in this relation as the set of *admissible* problem types.

6.3.5. FORMAL DESCRIPTION

We here pull together the concepts described thus far and give a formal description of the components of GSPS. We observe that this description, while necessary for translation to a usable computer implemented software package, is not necessary for comprehension of the overall conception which is adequately summarized in the next subsection.

Let E and M denote, respectively, sets of identifiers of all admissible epistemological and methodological criteria, and let J denote the set of identifiers of all *admissible system types.* Then:

$$J \subset E \times M$$

Let S be the set of all *particular admissible systems.* Then J imposes a partition on S:

$$\pi_J = \{S_1, S_2, \ldots, S_i, \ldots\}$$

where S_i is the set of all particular admissible systems of type i. Let $s_{i,j} \in S$. Then $s_{i,j}$ is a particular system of type i, identified by j.

Let $\mathfrak{R}$ denote the set of all *requirement types* defined within the context of the problem solver and let $\mathfrak{P}(\mathfrak{R})$ refer to the power set of $\mathfrak{R}$. Let R denote the set of all *particular requirements* which GSPS recognizes and let $\mathfrak{P}(R)$ denote the power set of R.

We distinguish two sets, P and P', such that:

$$P \subset P' \subset J \times J \times \mathfrak{P}(\mathfrak{R})$$

Set P' consists of all problem types which are *identifiable*; set P contains problem types which are both identifiable and *admissible* in the sense that they can be solved by methodological tools available in GSPS. Each identifiable problem type is thus a triple

$$(i, t, r) \in P'$$

where i and t denote, respectively, identifiers of the initial and terminal system types involved in the problem type, and r stands for a set of requirement types.

To simplify utilization of problem types in further definitions we define problem type identifiers through an assignment function f which assigns a different unique integer from $\{1, 2, \ldots, \#P'\}$ to each identifiable problem type. Thus we denote each identifiable problem type (i, t, r) by:

$$q = f(i, t, r)$$

In addition to those admissible problem types for which solution procedures exist, certain identifiable problems will be solvable as sequences of admissible problems. Let $Q = \boldsymbol{Pr}_{J \times J} P$ denote the projection of P into $J \times J$. Q thus represents all pairs of system types which participate in admissible problem types. Let Q_T denote the transitive closure of Q. Then, there is a set:

$$P_A \subset (Q_T - Q) \times \mathfrak{P}(\mathfrak{R})$$

which consists of problem types that are not admissible (cannot be directly solved by the available methodological tools) but can potentially be solved through appropriate sequences of admissible problem types. Problem types in set P_A will be called *potentially admissible.*

GSPS accepts as inputs particular problems, that is, problems which always involve particular systems and particular requirements. Outputs of GSPS, which represent solutions to particular problems, are of one of two kinds: either a particular terminal system of a type designated by the problem, or some property of the relation between two particular systems given in the problem. Therefore, particular problems come either from the set $S \times J \times \mathfrak{P}(R)$ or from the set $S \times S \times \mathfrak{P}(R)$.

Let PP be the set of all admissible particular problems and let PP_V and PP_w be two disjoint subsets of PP such that $PP_v \cup PP_w = PP$. Let $p_{q,v} \in PP_V$ and $p_{q,w} \in PP_w$. Then $p_{q,v}$ stands for the triple $(s_{i,j}, t, r_u)$ and we define $p_{q,v}$ to be a particular problem of type (i, t, r) identified by index q as introduced previously; r_u stands for the set of particular requirements of types identifed by r; v is an identifier of the pair (j, u). Similarly $p_{q,w}$ stands for the triple $(s_{i,j}, s_{t,k}, r_u)$ and we define $p_{q,w}$ to be again a particular problem of type (i, t, r) identified by q; w is an identifier of the triple (j, k, u). Elements of PP_v thus represent *problems of the first kind* and elements of PP_w represent *problems of the second kind*.

Problems of the first kind have the following canonical formulation. Given a particular initial system $s_{i,j}$, determine a terminal system of type t such that the requirements r_u are satisfied. A solution of a problem of the first kind is thus a particular system of type t.

Problems of the second kind have the following canonical formulation. Given a particular initial system $s_{i,j}$ and a particular terminal system $s_{t,k}$, determine some property, specified by the requirements r_u, of the relationship between $s_{i,j}$ and $s_{t,k}$.

The final aspects necessary in the specification of GSPS are components which identify the association of problem solving procedures with systems problems. The aspects which have been defined so far derive from specific and object-oriented research while the following aspects essentially tie these concepts to research in abstract disciplines.

Let F denote a set of fundamental procedures which are available to GSPS, and which involve manipulation of context-independent constructs. Let G denote the set of all sequences of fundamental procedures taken from F and let $T \subset G$ be a set of meaningful sequences of fundamental procedures in the sense that they are applicable to solving admissible problems. The meaningful sequences will be referred to as *methodological tools.* The association of the individual methodological tools with problem types can be described by a relation $A \subset T \times (P \times \{v, w\})$, where the identifiers v, w specify whether the problem is of the first or second kind. Similarly, the association of methodological tools with particular problems can be described by a relation $B \subset T \times (PP \times \{v, w\})$.

6.3.6. SUMMARY

The main concepts involved in the definition of GSPS are thus:

- Sets J, $\mathfrak{R}$, and P which identify system types, requirement types, and problem types; problem types are a subset of

  | system types | × | system types | × | requirement types |

  These sets arise from consideration of broad classes of contexts, but their definition is not dependent on given specific contexts.
- Sets (S, R, and PP) of particular systems, particular requirements, and particular problems; these sets are generally defined through interaction with context, but may also be independently specified.
- Sets F and T of fundamental procedures and methodological tools, respectively; these sets primarily derive from research in context-independent languages.
- Relations A and B which effect the connection between object-oriented and abstracted languages.

This situation, incorporating the concepts we have defined and representing GSPS as an integrative and mediating language, is depicted in Figure 10.

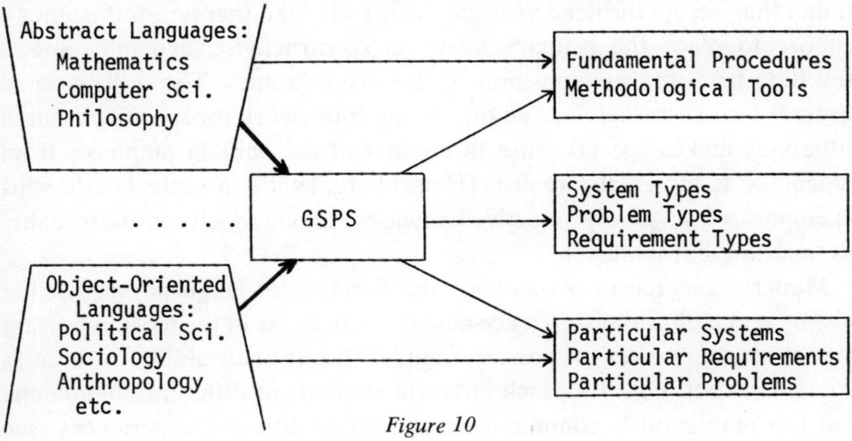

Figure 10

As we have described general systems research and developed GSPS they especially represent a framework through which object-oriented (or 'real-world') research can effectively make use of the simpler – but generally more precise and powerful – results of abstract research. Regarding this abstract/concrete interface the fundamental distinction in use between this framework and, say, the 'application to real-world problems' of particular techniques or results – such as linear programming and the simplex algorithm, or differential equation modelling and the use of transform methods – lies in the fact that it does constitute an integrative framework and that it strives for the most general formulation of problems, consciously attempting to overcome difficulties inherent in uncritical use of various assumptions.

With respect to the 'intractable' problems which are characteristic of much modern attention it is reasonable to argue, as did Weaver thirty years ago (see Chapter 2), that their complexity demands the use of investigative teams which represent expertise along a wide spectrum of knowledge. GSPS as a common point of focus and reference frame should be able to effectively deal with difficulties in communication and contribute to the joint working through of problems by those with expertise in specific object-disciplines, mathematics, philosophy and computer science.

6.4. GSPS as an Interactive Framework

The major emphasis in the development of GSPS has been on the acceptance of pragmatist and process-oriented conceptions which recognize merits in both objectivist and subjectivist epistemological positions but which – rather than accept the need to argue exclusively for either one or the other – choose to place the primary focus on constructivist, dynamic aspects resulting from the juxtaposition of the two extremes. The utilization of general systems research as an organizing framework for specific research situations makes special sense in context of the shift in emphasis from 'dogmatic' to 'pragmatic' realism (Heisenberg, 1958), in context of the shift in emphasis from purely objective knowledge to one which considers reality as 'resisting but malleable'.

Modern research in psychology, the function of language, and in the history and philosophy of science has shown that one of the most important aspects of the 'power' of powerful models lies in their ability to serve as communication links by which different students of different phenomena can less ambiguously communicate and relate differing experiences (see

Chapter 2; also, Winch, 1958; and Kuhn, 1962). This communication effectively contributes to a raised consciousness which can develop to an integration and acceptance into the general linguistic and conceptual structures of those who investigate various phenomena and thus in turn to new and more effective ways of perceiving and interacting with reality.

No matter how narrow the area of specific research which a specialist studies, the questions asked and the directions in which answers are searched for obviously must affect the answers which are found. Beyond this, it is often factors peculiar to individual researchers which play the ultimately major role in enrichment of knowledge regarding objects of study. The existence of a viable framework within which to embed diverse results facilitates recognition of the contributions of various perspectives and also facilitates critical and comparative evaluation of the results of these different emphases. The roles are particularly important for the non-simple areas of study such as are normally chosen as objects from within the social sciences. (For consideration of these issues from a slightly different perspective, see Cavallo and Conklin, 1977.)

We have considered the critical emphasis which must be given to methodological concerns when dealing with non-static conceptions of knowledge, an emphasis which has naturally played a major role in the study of humanistic and social systems. What has been lacking and recognized as needed (Boudon, 1970), and what general systems methodology is able to provide, is a conceptual framework which gives an overall 'view of the *total system of* [*methodological*] *tools* of which each special procedure is only a dependent part.'

This is not to suggest that GSPS constitutes an adequate investigative mechanism of itself, but that it *requires* interaction with expertise and object-knowledge which can only come from in-depth study and relation to specific phenomena. The development of general systems research allows this expertise and knowledge, which is related to specific research areas, to proceed and to provide the primary focus of study rather than allow diversionary consideration of information or data-processing methods to draw attention from what should be the true subject of the focused research. In this regard, the most meaningful use of GSPS is likely to be made in symbiotic relationships with investigators from specific disciplines and research areas, providing an organized mechanism for continuous two-way interaction between abstraction and interpretation. Figure 11 illustrates a representation of this interaction and of GSPS as an aid for investigative research.

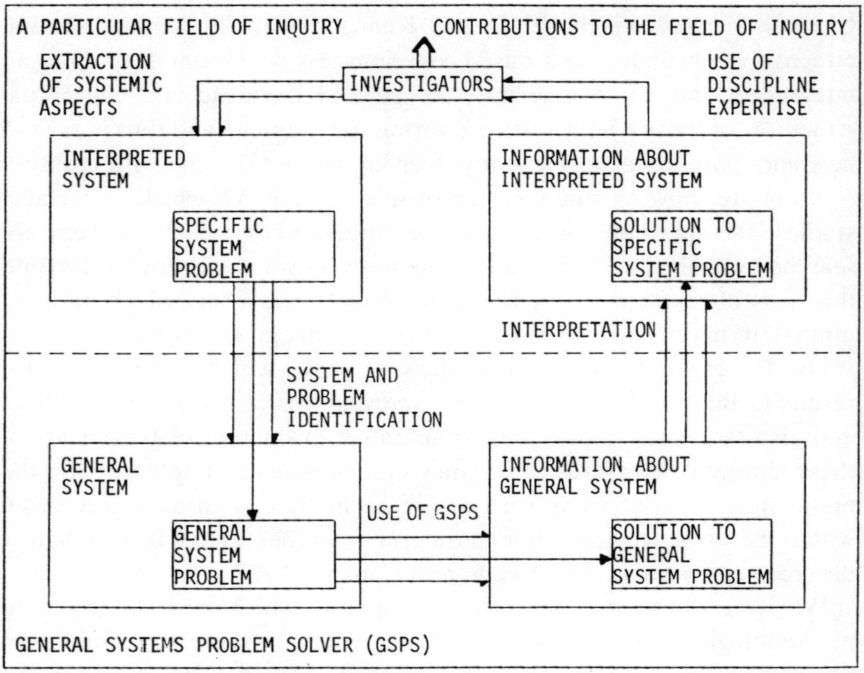

Figure 11

Fundamental to this characterization is the recognition that general systems research is not likely to turn complex and intractable overall problems into ones which are suddenly 'solvable.' It does appear likely, however, that an organized and general investigative approach can significantly contribute to an advanced understanding of these complex areas of research.

Reflecting the framework we have described, GSPS can be considered to operate on two levels:

1. [Represented by the inner rectangles.] The investigator is familiar enough with the basic language of GSPS to be able to formulate an interpretation of a system problem within his own discipline. In this case the investigator, or user, maps the interpretation to a GSPS formulation. GSPS solves the general systems problem and maps the solution to the interpreted system.
2. [Outer rectangles.] Many systems investigations are of sufficient complexity that the investigator can make meaningful use of more information than that provided by the solution to a particular system prob-

lem. In this case, also, general aspects of overall problems will indicate a transformation to a general system. Based on the information supplied through this transformation, GSPS can translate new information about the general system – utilizing the epistemological-level framework – back to the interpreted system. The investigator is thus given new knowledge about the interpreted system which represents enrichment by differing and complementary viewponts embodied in and transferred from the systems framework conceptions.

Part 2: General Systems Problem Solving and the Study of Domestic Conflict

6.5. Introduction

The intention of this part is to exhibit the capability which GSPS provides for organizing an overall investigation, reporting on a utilization – as an investigative aid in the study of conflict within nations – of the framework which has been developed to this point. There are several reasons that this specific social science research area has been chosen to demonstrate the applicability of the framework. The organization of the material which follows centers around three of the most important of these reasons:

- The study of domestic conflict is an important research area within political science which has received a reasonable amount of attention in recent years, including the compilation of extensive data.
- This attention has proven reasonably fruitful in contributing to the understanding of patterns and relations underlying conflict phenomena.
- It is undoubtedly the case that we are still reasonably far from a satisfactory understanding of the major issues involved.

The approach taken in the following sections, after reviewing past work in the area, is to run through the use of GSPS – both conceptually and operationally – from the stage of system definition to that of structure determination indicating the manner in which systems problems naturally arise and how solutions to these problems provide basic insight into the area of investigation. In addition to providing information relevant at all epistemological levels of system definition the results will be seen to provide a 'theoretical' structure for the underlying object system. This latter contribution constitutes what Singer (1971) refers to as 'explanatory knowledge.'

6.6. The Need for General Operational Methods

Exploration of relations among domestic conflict indicators has been undertaken and reported for more than a decade (e.g., Rummel, 1963, 1965; Tanter, 1966; Firestone and McCormick, 1972; Banks, 1972; Gurr and Bishop, 1976; Hibbs, 1973). This research reflects a clear example of a desire to go beyond or augment, on the one hand, verbal and interpretive studies of individual cases and, on the other, grand sweeping theories of why nations behave the way they do. The studies represent an effort to proceed in a manner representing systematic integration of more than a few instances of phenomena under consideration, that is, to proceed in a manner which uses systematic methods for the ordering and surveying of human experience.

All of the research under consideration utilizes data drawn from similar sources – for example, newspaper accounts, *Facts on File*, *Deadline Data on World Affairs* – and fits into the category classified as *events research* (Azar and Ben-Dak, 1975). The compilation of such data – and its being made generally available for further research – represents a significant achievement in its own right as a first step toward a more complete understanding of political phenomena. This non-experimental research area is thus one in which the tension between theoretical and empirical orientations particularly highlights the need for an operational investigative framework.

Essentially all of the past research in this area has centered around the linear system based statistical tools of correlation analysis, regression analysis and factor analysis. Indeed, many social scientists – while often recognizing the unsatisfactory nature of the situation (Singer, 1976) – consider these tools as constituting the only techniques available to workers in this field.

Over the years Rummel has probably been the most enthusiastic proponent of factor analysis (Rummel, 1970, 1967, 1965, 1976) and it is interesting to consider his investigations from our perspective. Rummel (1965) indicates a basic affinity with systemic emphases, arguing that:

> With its accent on mathematics and the interdependencies of elements within a system, general systems theory has been a stimulus to social theorizing. It changed the focus from phenomena to patterns and relations, and has had a purging and heuristic effect on current social thought and research (Rummel, 1965).

Rummel goes on, however, to argue that general systems theory has stagnated and that this has occurred for two intricately related reasons. The first of these is that, 'as used by social scientists,' general systems theory has not gone beyond the provision of a conceptual and 'verbal edifice.' In this regard Rummel possibly underemphasizes the importance of the shifted emphases which general systems and cybernetic foci have represented (see, e.g., Deutsch, 1963). His second argument, however – that general systems theory was 'ungrounded in empirical data and operational concerns' and that it did not provide 'a developed set of empirical methods' – relates directly to the issues which have motivated development of this framework, issues that Boulding had referred to ten years earlier (see Chapter 3), and that Melcher would refer to ten years later regarding the lack of usefulness of general systems concepts (see Chapter 2).

With respect to Rummel's two arguments, it is most likely the case that it has been the lack of empirical methods – or, more accurately, the failure to relate empirically significant ideas and methods to social science research – that has been responsible for the primarily conceptual utilization of general systems research by social scientists (and in this regard we recognize that conceptual refers to both purely verbal theories and to purely symbolic or mathematical explications which do not give consideration to investigative utility).

Rummel has thus been led to total reliance on product moment correlation, multiple regression, and principle axes factor analysis to supply a model with which to deal with empirical relations. This model has been subsequently used to formulate a 'field theory' of social action which is essentially a system of linear equations deriving from a basic 'axiom' that the relation between two nations which defines their behavior is a 'linear vector function' of attributes which the nations possess.

Rummel does, in fact, recognize the restrictiveness of the assumptions involved in the use of his models but evidently felt that there was no alternative to their use. This recognition is also made – but also essentially ignored – in other basic methodological social science works, for example: Hibbs, 1973; Heise, 1975; Cortés, et al., 1974; and Blalock, 1961, who states that 'it is the regression coefficients which give us the laws of science'' (p. 51). A major purpose of this framework is to extend this methodological scope.

6.7. Past Work

A major orientation of past studies of domestic conflict has been the attempt to determine, through the use of factor analysis, a number of 'dimensions' which is lower than the number of attributes by which the conflict system is defined, such that these dimensions adequately explain the overall system. 'Adequate explanation' must be defined in terms of the approach taken and in this context refers to accounting for an acceptably high proportion of the overall variance. In terms of the framework which we have presented this goal can be considered as the determination of a structure system which is derived from the overall conflict situation.

Rummel (1963), for example, investigates nine domestic conflict measures: Assassinations, General Strikes, Guerrilla Warfare, Major Government Crises, Purges, Riots, Revolutions, Anti-Government Demonstrations, Domestic Number Killed. Values of these attributes were collected over seventy-seven nations and summed for the years 1955-1957. The values were then transformed by log transformation or by lumping (or scaling) according to a geometric progression.

Through factor analysis of this data Rummel then determined three conflict behavior dimensions: a *turmoil* dimension, a *revolutionary* dimension, and a *subversion* dimension. Given the assignments which Rummel makes through interpretation of the factor loadings we may consider his overall system to have been decomposed into three subsystems as pictured in Figure 12.

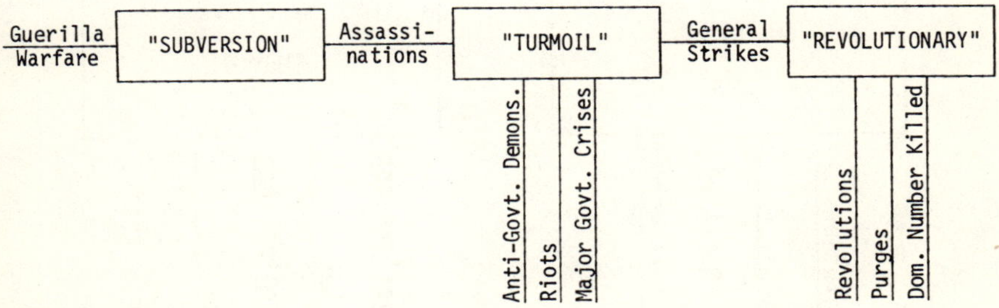

Figure 12

General Strikes and Assassinations are characterized as coupling variables between subsystems since it is not clear from Rummel's explication which 'dimension' they should belong to (that is, the factor loadings are such that neither of the two load significantly on only one factor). While Rummel's interpretations – as must any, by the very definition of 'interpretation' – contain what may be considered subjective or arbitrary elements, and while there are arbitrary elements inherent to the use of factor analysis itself (Armstrong, 1967; Einhorn, 1972; Armstrong and Soelberg, 1968; Hazlewood, 1976; Digman, 1966), the determination obviously represents a non-trivial 'survey and ordering' of past experience. Rather than replicate consideration of limitations bound to the particular method we merely observe that the same questions addressed by Rummel's factor analysis can be addressed from a less restrictive perspective, and that there are meaningful questions which have not been addressed but which can be approached through the use of the general systems framework.

Before considering these we describe other work which has attempted to develop or generalize Rummel's findings. Tanter (1966) has reproduced Rummel's investigation of data for 1955-1957, with data for 1958-1960, for eighty-three nations, with the objective of determining the stability of findings from one period to another (each of these studies has also considered foreign conflict but as the interest here is primarily methodological we limit consideration to that of domestic conflict).

Tanter felt that two of Rummel's dimensions – revolutionary and subversive – should more properly be merged for the 1958-1960 data into one which he labels internal war. He is also much more definite regarding subsystem separation, declaring that one subset of attributes defines one dimension while its complement defines another. Tanter's structure system is represented in Figure 13.

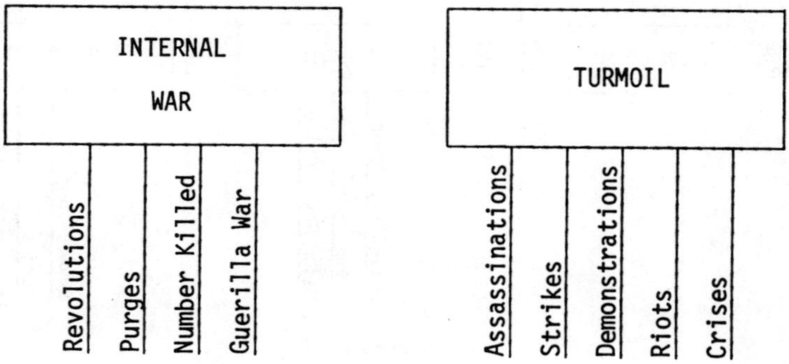

Figure 13

Although it is possible to agree with certain overall similarities in the two studies, Tanter's interpretation of his results tends to underemphasize differences between his effort and that of Rummel. For example, although assassinations and general strikes were found by Rummel to load reasonably highly on the turmoil dimension, each of them loaded more heavily on other (and different) dimensions. Thus Tanter's contention that 'the internal war dimension combines the 1955-1957 subversion and revolutionary dimensions; the turmoil dimension, however, is found in both periods' appears unwarranted or, at least, requires more careful qualification. Tanter's effort was valuable, however, in that the comparison of two succeeding periods can be considered as a preliminary concern with the introduction of dynamics into the study of domestic conflict.

Banks (1972) extended this consideration by performing similar analyses (on only eight of the indicators – domestic number killed was dropped) utilizing data from 1919-1966 excluding the war years 1940-1945. As Banks' intentions are to compare results in different periods he only uses data from the fifty-one countries continually independent over the period in question.

Banks' procedure was to sum the data for seven-year periods, use the same geometric scaling as in Rummel and Tanter and assign the weighted aggregates to the six midpoint years 1922, 1929, 1936, 1949, 1956, 1963. Banks ran six R-mode factor analyses and found that 'in every case, a three-factor structure emerged, the components of which could roughly be interpreted as delineating Rummel's turmoil, revolutionary and subversive dimensions. The structure was, however, far from stable through time particularly with respect to the second and third factors.'

Banks does not give the results on the six factor analyses but rather groups the pre-war data into one set, the post-war into another, and runs three-dimensional P-mode factor analyses on these as well as on the combined 1922-1963 data. The structures so determined are listed in Figure 14.

6.8. Systems Problems Related to Source, Data, and Behavior Systems

Contemplation of Banks' results in conjunction with those of Rummel and Tanter points to the desirability of reconsideration of the data from a more comprehensive perspective, one which utilizes data for a significant time period and for a large number of countries and which at the same time avoids the loss of information inherent in the procedure – used in all the above studies – of summing observations over periods of years. Especially in a seven-year sum this represents potentially severe distortion of actual interaction patterns.

In the sequel we utilize data compiled by Banks at the Center for Comparative Political Research at the State University of New York at Binghamton (made available through the Inter-University Consortium for Political and Social Research at Ann Arbor, Michigan).

6.8.1. SOURCE SYSTEM AND DATA SYSTEM

In the overall investigation which we are describing the object is thus 'conflict within nations.' Definition of the object system requires the speci-

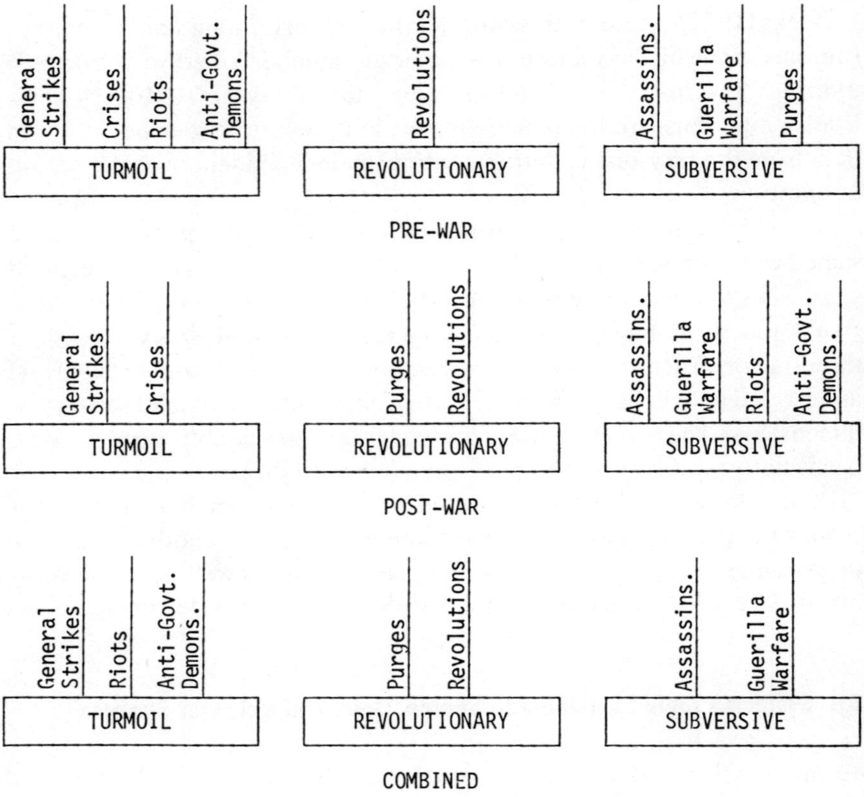

Figure 14

fication of attributes and a set of appearances for each attribute. In this case the attributes have been determined in consideration of the past studies and it was decided to utilize the set consisting of the six elements: General Strikes, Major Government Crises, Purges, Riots, Revolutions, Anti-Government Demonstrations. Assassinations and Guerrilla Warfare were not used since appearances of these attributes are so highly susceptible to non-systemic effects. Definition of the six attributes are as given by Rummel (1963):

Number of general strikes: any strike of 1,000 or more industrial or service workers that involves more than one employer and that is aimed at national government policies or authority.

Number of major government crises: any rapidly developing situation that threatens to bring the downfall of the present regime – excluding situations of revolt aimed at such an overthrow.

Number of purges: any systematic elimination by jailing or execution of political opposition within the ranks of the regime or the opposition.

Number of riots: any violent demonstration or clash of more than 100 citizens involving the use of physical force.

Number of revolutions: any illegal or forced change in the top governmental elite, any attempt at such a change, or any successful or unsuccessful armed rebellion whose aim is independence from the central government.

Number of anti-government demonstrations: any peaceful public gathering of at least 100 people for the primary purpose of displaying or voicing their opposition to government policies or authority, excluding those demonstrations of a distinctly anti-foreign nature.

The appearances of each attribute are the recorded events as given in the above named data bank.

For the definition of the parameter space it was decided to use the fifty-one countries used in the Banks study and to consider observations on each country for the post World War II period consisting of the years 1946-1975. The parameter space thus consists of $51 \times 30 = 1530$ points which are linearly ordered with breaks at intervals of 30.

For definition of the general image system the only significant decision involves the mappings from the sets of appearances to the sets of recognized potential states of the abstract variables. With respect to this decision it seemed necessary to take into account two basic considerations. These were, first, the limitations inherent in cross-sectional data which force simultaneous consideration of data collected in diverse settings and, second, the nature and complexity of the overall situation, coupled with difficulties associated with the collection of data, which make it more meaningful to utilize a low resolution level and search primarily for patterns which the data exhibit. In these regards the fact that a given event will have different significance in different social and cultural settings was taken account of by associating each attribute with a binary variable where the state was determined by whether or not the appearance for the country in question was above or below the mean for that country over the thirty-year period. In this manner the variable state reflects the relative weight of events in the context history of each nation rather than in comparison to the frequency of events in all nations. The data system was thus represented by a data array consisting of a 6×1530 binary matrix with delimiters set at thirty-column intervals. Table 3 lists the fifty-one countries used in the study.

Table 3

Afghanistan	1	10	El Salvador	27	350
Bolivia	2	100	Finland	28	380
South Africa	3	1040	France	29	390
Spain	4	1060	Argentinia	30	40
Norway	5	1091	Greece	31	450
Sweden	6	1092	Guatemala	32	460
Switzerland	7	1100	Haiti	33	490
Thailand	8	1130	Australia	34	50
Turkey	9	1170	Honduras	35	500
USSR	10	1190	Iran	36	540
Brazil	11	120	Italy	37	570
UK	12	1210	Hungary	38	62
US	13	1220	Liberia	39	690
Uruguay	14	1240	Luxembourg	40	720
Venezuela	15	1250	Belgium	41	80
Yugoslavia	16	1290	Mexico	42	810
Bulgaria	17	130	Netherlands	43	850
Canada	18	180	New Zealand	44	860
Albania	19	20	Nicaragua	45	870
Chile	20	220	Panama	46	910
China	21	230	Paraguay	47	920
Columbia	22	240	Peru	48	930
Costa Rica	23	270	Portugal	49	960
Cuba	24	280	Rumania	50	970
Denmark	25	320	Saudi Arabia	51	990
Ecuador	26	340			

Associated with each country are two numbers. The first is a number from one to fifty-one which indicates the thirty-column interval of the data array representing data for that country. The second is a code number representing that country and is as given in the original data bank. The latter numbers are used in Appendices 1 and 2 to identify the countries. In Appendix 1 the raw data is given for each country and Appendix 2 contains the fifty-one 6×30 binary matrices which were catenated to form the overall data array. The rows of the matrices represent the attributes and variables in the order previously given, that is:

Row 1 : v_1 : General Strikes
Row 2 : v_2 : Major Government Crises
Row 3 : v_3 : Purges

Row 4 : v_4 : Riots
Row 5 : v_5 : Revolutions
Row 6 : v_6 : Anti-Government Demonstrations

As an example, C1040 and C1040A are reproduced here from Appendices 1 and 2 respectively:

<u>C1040</u>

2	1	0	0	0	1	0	0	0	0	0	0	0	0	0	0	0	0	0	0	0	0	0	0	0	0	1	1	0	0
0	0	0	0	0	0	1	0	0	0	0	0	0	0	1	0	0	0	0	0	0	0	0	1	2	0	0	0	0	0
1	0	0	1	1	0	2	1	0	0	1	0	1	0	1	0	2	1	0	1	0	0	0	0	0	3	2	0	0	2
0	0	0	5	3	1	14	2	3	0	2	2	1	2	8	3	1	2	0	0	0	2	0	0	2	0	5	2	4	0
0	0	0	0	0	0	0	0	0	0	0	0	0	0	0	0	0	0	0	0	0	0	0	0	0	0	0	0	0	0
2	0	0	1	0	0	9	1	0	1	1	3	3	1	2	0	3	0	0	0	0	0	0	1	3	0	3	0	0	0

<u>C1040A</u>

1	1	0	0	0	1	0	0	0	0	0	0	0	0	0	0	0	0	0	0	0	0	0	0	0	0	1	1	0	0
0	0	0	0	0	0	1	0	0	0	0	0	0	0	1	0	0	0	0	0	0	0	0	1	1	0	0	0	0	0
1	0	0	1	1	0	1	1	0	0	1	0	1	0	1	0	1	1	0	1	0	0	0	0	0	1	1	0	0	1
0	0	0	1	1	0	1	0	1	0	0	0	0	0	1	1	0	0	0	0	0	0	0	0	0	0	1	0	1	0
0	0	0	0	0	0	0	0	0	0	0	0	0	0	0	0	0	0	0	0	0	0	0	0	0	0	0	0	0	0
1	0	0	0	0	0	1	0	0	0	0	1	1	0	1	0	1	0	0	0	0	0	0	1	0	1	0	0	0	0

The eighth column of C1040 indicates one instance of Purge activity, two Riots, and one Anti-Government Demonstration reported for South Africa for the year 1953. Since the two Riots and one Anti-Government Demonstration are lower than the average over the thirty-year period, the respective places in the eighth column of C1040A are assigned zeros, while evidence of the Purge activity is retained.

6.8.2. BEHAVIOR SYSTEMS

In the context of an overall investigation, using GSPS as an organizing framework as we are describing here, the first class of problems which are relevant relates to the determination of behavior systems from data systems. These two types of systems constitute direct analogues of 'existential knowledge' and 'correlational knowledge' – terms used by Singer (1971) in a more conceptual investigation of the stages of knowledge acquisition involved in the epistemological process as related to the investigation of social systems. The essence of the methodologically pluralistic perspective of GSPS allows and motivates the specification of different problems related to this transition and thus the determination of different behavior systems to answer different questions which arise directly from the consideration of the specific system.

Memoryless Behavior

The most direct problem is the determination of the memoryless behavior from the data array, that is, the specification of the behavior representing interaction among only current values of the basic variables. Singer's classification of knowledge at this level as 'correlational' refers to a knowledge of 'which events or conditions go together in time or space.' This notion is completely captured in the frameworks – as described in Chapter 4 – through the definition of behavior systems, though 'correlational knowledge' implies a methodological reliance on correlation coefficients (Singer, in fact, illustrates his explanation of knowledge at this level through a discussion involving correlation coefficients). It is worth re-emphasizing that the information theoretic measures which we primarily utilize in this framework provide a measure of the degree to 'which events go together in time' rather than a measure of the degree to which different events are linearly related – as is the case with correlation coefficients. The use of the information theoretic approach has the further advantage of providing a measure upon which a calculus or set of operations may be performed. This fact is particularly useful in context of the overall investigation – in context of the desire to go from correlational to 'explanatory knowledge' (Singer), that is, to derive meaningful structure systems from the behavioral representation. This sequence is illustrated in detail in the next section (see also Krippendorf, 1978, and Broekstra, 1978, for alternate detailed considerations of the development of the measures in the context of 'explanatory knowledge' generation).

Formulated as a problem in GSPS, the determination of behavior is represented by:
– a given particular initial (data) system
– a given terminal system *type* (behavior)
– constraints, specified as the request for memoryless behavior
The solution to the problem consists of a list of the samples which occur, along with the frequency-defined probability of each sample as described in Chapter 4. The resulting behavior, listed as six-digit binary numerals with associated probabilities, is given in Appendix 3. As an example, the numeral 010101 with associated probability .009 indicates that over the whole

data array the particular combination – Major Government Crises, Riots, and Anti-Government Demonstrations – represented roughly one percent of the sample.

A major reason for the determination of this behavior is as a solution of an *admissible* problem which forms part of a two-problem sequence which together constitute the *potentially admissible* problem of the determination of structure directly from the data array. (This problem is explicitly considered since it is the direct analogue of the work of Rummel, Tanter and Banks reported earlier. In the latter studies the steps are not always made explicit but are embedded in the use of factor analysis.)

The data given in Appendix 3 represent information on all occurrences and interactions within each nation over the full thirty-year period. Investigative areas such as we are describing are especially characterized by the lack of a well-defined and accepted set of general ordering principles. In these cases maximum utilization of general systems methodological research implies interpretive interaction with the user (that is, interaction between general and specific system) involving investigations of an exploratory nature. Such exploratory considerations can fruitfully be made at all levels.

For example, even though the behavior system may not represent a primary investigative goal with respect to the object-system, its determination is in general worth examining for the detection of patterns or interactions of interest. This is especially true in context of the overall framework which embodies a spectrum of techniques geared to the extension and amplification of the intuitive facilities of the investigator. In the form in which it is given in Appendix 3 the behavior is not particularly suitable for such consideration. For example, Purge activity could appear in as many as thirty-two of the sixty-four possible samples and it would be difficult to pose questions directly related to this aspect of conflict.

This situation is, however, easily remedied through the use of an adaptation of diagrams introduced by Marquand in 1881 (to aid in the investigation of logical propositions and set intersections). Figure 15 represents the diagram for the six variables where, for example, the outlined entry representing $\sim v_1, v_2, \sim v_3, v_4, \sim v_5, v_6$ and containing the probability .009 represents the same sample which was discussed earlier ($\sim v_1$ represents $v_1 = 0$). The last column and last row give the sums for the rows and columns. With the data presented in this form it is easy to sum columns or rows to get overall information such as total percentage of occurrences involving Riots (v_4; columns 5, 6, 7, 8) or those involving Riots but not Anti-Government Demonstrations (columns 5, 7). Such exercises will of course have more or less importance depending on the general state of knowledge regarding the area under investigation (for further discussion of this information see Cavallo and Ziegenhagen, 1978). As the framework also involves the association of methodological procedures – such as clustering techniques, pattern recognition algorithms, fuzzy algorithms – to various systems problems, interaction with the user will often suggest systems problems which involve these procedures. An example of a problem which naturally arises in the context of this investigation is the determination of idiosyncratic or anomalous system states, where such states may be characterized by different possible criteria. The detection of such anomalies is especially important in the investigation of complex areas for which theoretical formulations – if they are possible – do not exist.

A characterization of anomalous states which is relevant to this investigation considers meaningful (or especially large) discrepancies between the probabilities associated with a given state (where state refers to a particular interaction or co-occurrence phenomenon) and those states which are most similar to that state. Such similar states are called 'immediate neighbors' and are defined as those resulting from a change (in the absence or presence) of only one of the variables. Which discrepancies are 'meaningful' are in general not specifiable with respect to an absolute criterion but must be determined within the context of the investigation. (In terms of GSPS, this problem is formulable as the determination of a terminal behavior system given an initial behavior system, where states in the initial behavior system are assigned values based on the probabilities and where the terminal system assignments should be based on consideration of each state's neighbors.)

One procedure which was adaptable for solution of this problem has been developed by Barto and Davis (1977) in the context of Consensus-Competition models. Since an adequate description of the procedure (which considers the six-component binary numerals as elements of the dyadic group $\{0, 1\}^6$ with addition of two six-tuples defined by digit by digit exclusive-or) would involve somewhat diversionary consideration of mathematical concepts we will not describe the procedure in detail here. In context of this application any given procedure is not primarily relevant, the major point being the ability of GSPS to marshal significant results or procedures for utilization in specific systems investigations and to make them available to context oriented researchers, removing the necessity of involved study of abstract language

			v_4				v_4				
			v_5		v_5		v_5		v_5		
			v_6	v_6	v_6	v_6	v_6	v_6	v_6	v_6	
v_1	v_2	v_3	.4582	.0327	.0229	.0046	.0569	.0216	.0105	.0333	.6105
		v_3	.0732	.0131	.0163	.0013	.0163	.0046	.0026	.0039	.1314
	v_2	v_3	.0601	.0118	.0098	.0020	.0144	.0092	.0026	.0052	.1150
		v_3	.0131	.0052	.0065	.0020	.0039	.0039	.0098	.0046	.0490
v_1	v_2	v_3	.0203	.0013	.0007	.0000	.0052	.0065	.0007	.0007	.0353
		v_3	.0059	.0020	.0000	.0007	.0039	.0033	.0007	.0007	.0170
	v_2	v_3	.0059	.0013	.0013	.0013	.0052	.0065	.0013	.0020	.0248
		v_3	.0052	.0007	.0020	.0000	.0013	.0020	.0013	.0046	.0170
			.6418	.0680	.0595	.0118	.1072	.0575	.0294	.0248	1.0000

Riots (.2189)

Riots without Anti-Government Demonstrations (.1366)

v_1 = General Strikes

v_2 = Major Government Crises

v_3 = Purges

Figure 15

considerations. This example is especially illustrative since its results, once found, are independent of abstract ideas and may be easily and completely evaluated in context of the specific system.

Use of the procedure on this data (to detect anomalous states) identified four states with probabilities significantly different than a weighted combination of their neighbors. Three of them were: General Strikes and Riots, General Strikes and Purges, General Strikes and Government Crises. All three occurred less frequently than their neighbors. A check of the neighboring states indicates that these three potential anomalies are contextually reasonable in light of the high incidence of the neighboring state consisting of just General Strikes. That is, it is reasonable to expect a significant divergence in frequency between the occurrence of only General Strikes, on the one hand, and the occurrence of General Strikes in conjunction with another attribute on the other.

The fourth anomalous state, however, is interesting in that it involves a large amount of conflict activity and occurs significantly more often than its neighboring states. The state was that indicating the co-occurrence of Government Crises, Purges, Riots and Revolutionary Activity. This state occurred fifteen times. The neighboring states, along with their frequency of occurrence, are given in Table 4. The high incidence of Riots and Revolutions in the presence of combined Crisis and Purge activity (011110), compared to that in the presence of Crises (011100) or Purge (001110) activity alone, appears to warrant further consideration on the part of scholars in this area (see Cavallo and Ziegenhagen, 1978).

The unique nature of that particular high conflict state is further emphasized by considering that, while sixty-one of the sixty-four possible states occur at least once, 87 percent of the occurrences are accounted for by eighteen states, with only one other of the eighteen states – Government Crises, Purges and Revolutions – involving as many as three dimensions. The average number of occurrences for other states involving four dimensions is less than four.

Table 4

BINARY REPRESENTATION	STATE	FREQUENCY
011110	Gov. Crises, Purges, Riots, Revs.	15
011100	Gov. Crises, Purges, Riots	6
001110	Purges, Riots, Revs.	4
010110	Gov. Crises, Riots, Revs.	4
011010	Gov. Crises, Purges, Revs.	10
011111	Gov. Crises, Purges, Riots, Revs., Anti-Gov. Dems.	7
111110	Gen. Strikes, Gov. Crises, Purges, Riots, Revs.	2

Memory Effects

The extent of past attempts at integration of dynamic effects in the study of domestic conflict has been limited to comparison of results of factor analyses over different aggregated periods (Banks, 1972; Tanter, 1966; Hibbs, 1973; Firestone and McCormick, 1972). In context of this framework these considerations may be considerably extended through the determination of generative behavior systems as described in Chapter 4.

A first determination of lagged or memory effects involves an evaluation of past values of variables and combinations of variables to determine which contribute most significantly to knowledge of the current state. As a problem in GSPS the initial system and terminal system type are as in the determination of memoryless behavior, but the requirement set involves determination of the best mask (sampling scheme).

Determination of the effects of past values of individual variables, of pairs, etc. indicated no most significant variable or combination, where past value of a given *single* variable, for example, could be considered especially significant if it contributed more to the reduction in uncertainty about the current state of the system than, say, past values of other *pairs* of variables. In general, the reduction in uncertainty was a strictly monotonic function of the number of past values used, though with respect to each number certain sampling variables of course contributed more than others.

For example, considering single variables, past values of the revolution dimension gave the most information about the current state (of the overall system) while general strikes contributed least. Interestingly, if pairs of past values are allowed, purge activity plus riots contributed more to knowledge of the current state (this is loosely equivalent to the ability to predict the next state) than any other pair, including those involving revolutions. This is a further illustration of the fundamental systems point regarding the inability to linearly combine particular pieces of information. Pairs involving any two of the three dimensions General Strikes, Government Crises, and Anti-Government Demonstrations contributed least information. These three as a set also were the least useful among all sets of three variables while Purges, Riots, and Revolutions gave the most information.

Because of this general lack of discrimination it was decided to use past values of all variables to process the data and, because of the intuitive significance of a single state involving one time period of all dimensions, to not consider further past values. A state of the system thus involves only and all current values of all attributes. The determined behavior is thus specifiable as a 64 × 64 state transition matrix. This matrix is reproduced in Appendix 4, where each of the column and row numbers represent the decimal equivalent plus one of the binary numeral indicator of a state. For example, the entry in the ninth row, fifth column indicates the number of transitions from the binary equivalent of eight to the binary equivalent of four, that is from 001000 to 000100. There were thus ten transitions from the conflict state involving just Purge activity to that involving just Riots.

Since the matrix in Appendix 4 represents all transitions which have occurred between conflict states for fifty-one countries in the post-war period, it represents close to all that can be said regarding single-step movement among these indicators. Consideration of the matrix indicates, however, that there are few context-independent generalizations which can be made regarding these transitions. At its worst, the lack of very striking indications may itself be considered worthwhile as it forces the researcher to face the problems in generalization which exist regarding the data, rather than lending an air of certainty to (basically false) conclusions through the spurious use of 'models.'

Regarding the perceived need of such simple models and what sense they make for complex systems, it is interesting to consider as an example the Keynesian model from economics. This model has been enviously and wistfully regarded by political scientists, and even referred to as a 'towering edifice' of the social sciences (Holt and Richardson, 1970). It is important here to compare the judgement of a Nobel laureate in economics, F. A. Hayek, who deems the Keynesian influence an 'unfortunate episode of... monetary history,' during which 'the whole Western world allowed itself to be led into [a] frightful dilemma [inflation plus unemployment].' Hayek continues that 'the Keynesian dream is gone even if its ghost will continue to plague politics [and we might add political scientists] for decades' (Hayek, 1975).

While Hayek's judgement may underestimate the pragmatist value which the Keynesian model had in context of an especially important period of economic history with pressing short-term needs, it would nevertheless appear to be folly to not heed his warning against the search for 'cheap and easy' solutions to complex problems and against the faith or credibility which should be accorded such solutions.

In terms of constructing a dynamic model of conflict phenomena, honest appraisal of the actual facts – as represented by the matrix in Appendix 4 – only underscores the difficulties inherent in such an endeavor. In light of such facts, the value of attempts to jump directly to 'theoretical' generalizations seems especially questionable. For example, there do not seem to be many valid conclusions that can be drawn through comparison of factor analytic structures determined at different points in time. The difficulties involved are only compounded when a factor analysis is run for a period which combines data for a number of years, as most of the dynamics may be lost in this aggregation. Period by period comparison of factor analyses is also rendered somewhat meaningless in light of the fact that all information is lost regarding change from specific system state to specific system state.

The situation here is a good illustration of the distinctions made by Weaver (1948; see Chapter 2) between simple and disorganizedly complex systems on the one hand and organizedly complex systems on the other. The essence of this distinction lies in the fact it is neither the case that the system is simple enough to only consider individual identities (for example, as in an historical study of the French Revolution) nor is it the case that the system is such that the only significant properties are emergent (or statistically defined) properties, where individual identities are basically irrelevant. We refer again to the statement by de Broglie (p. 92) and point out that we are clearly dealing with a situation in which the desire to construct a general and/or simple model is mitigated by the equally pressing demand to retain the ability to make significant statements about individual cases.

In this regard, there are at least certain general facts which are highlighted by the state-transition matrix. For example, of the twenty-two states involving Government Crises and Rioting, the only large number of transitions (seven) is to just Rioting, while Government Crises in conjunction with Anti-Government Demonstrations is followed most frequently by a next-state of no conflict activity. Meaningful evaluation of such information obviously requires that consideration be given to context-dependent past studies of elite conflict in conjunction with different types of mass disorder. Such studies have been oriented toward determination of the role that such conflict combinations are thought to play with respect to overall order in stable and in changing societies (Huntington, 1968; Edwards, 1927; Brinton, 1952; Hagopian, 1974), and these studies can be meaningfully augmented through interaction with GSPS.

Information such as is obtained from general systems methodological considerations can thus be given context through comparison with these studies. In the case of an extensive interactive study this will generate further systems problems – for example extensions of past consideration of certain variables, or the comparison of movement among different indicators in countries which have experienced revolutionary activity versus those which haven't. The general system determination also serves as a means by which hypothetical constructions deriving from individual studies can be evaluated in light of a more comprehensive consideration of data. As the focus of this chapter is mainly methodological and intended primarily to demonstrate the applicability and utility of GSPS, these more detailed determinations are not reported here but will form the basis of a further study which is more directly context oriented.

In this section we have considered behavior from both static and dynamic perspectives. In context of an overall investigation using GSPS this involves the second of the hierarchically organized epistemological systems levels and leads naturally to the consideration of structure systems, to an attempt to derive explanatory knowledge.

6.9. Systems Problems Related to Structure

We have described aspects of the major attention which has been given by investigators of domestic conflict to the factor analytic determination of structure. In this section we consider the determination of structure as a systems problem in GSPS involving transition from behavior to structure system. While it would be possible to consider factor analysis as a tool within GSPS to achieve this transition, our emphasis has been primarily on methodological processes which involve minimal assumptions and in this regard we utilize procedures which have been described in Chapter 5 to determine the confidence which may be associated with various structure systems.

It is possible to integrate into this problem – through the choice of particular requirements – various criteria with respect to which structure systems may be evaluated. Which criteria are chosen, and what significance should be given to the results will in general depend heavily on the contextual questions which are responsible for the generation of the systems problem. In this case we only look at the problem in regard to general criteria and do not give major consideration to context-related requirements.

We recall that the basic conception of a structure system is one which expresses an overall system in terms of subsystems and relations among them. The constraint imposed on such constructions is that some property of the overall system is tolerably accounted for by the structure system; that is, that with respect to some criterion an acceptably minimal amount of information regarding this criterion is lost by breaking the overall system into smaller parts. We have observed that one criterion which is close to the unadulterated behavior system – defined by a set of samples augmented by a probability distribution – is that which measures the difference between the probabilities associated with the samples in the overall system and those which are implied by different postulated structures. Since this criterion, as well as others, is used below we describe its use by a simple example. Consider the system with three binary variables described by Figure 1 of Section 5.3, which we may depict as Figure 16a. Figure 16b reproduces Figure 2 from Section 5.3 and gives the projections onto each two-dimensional subspace of the overall relation which defines the system, along with associated probabilities. Figure 16c gives the projections onto the one-dimensional spaces.

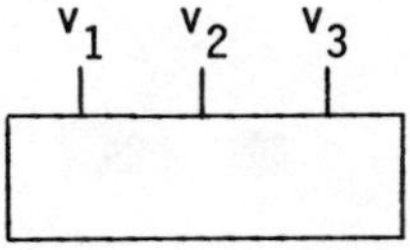

a

v_1	v_2	$p(v_1, v_2)$	v_2	v_3	$p(v_2, v_3)$	v_1	v_3	$p(v_1, v_3)$
0	0	.25	0	0	.02	0	0	.25
0	1	.25	0	1	.48	0	1	.25
1	0	.25	1	0	.48	1	0	.25
1	1	.25	1	1	.02	1	1	.25

b

v_1	$p(v_1)$	v_2	$p(v_2)$	v_3	$p(v_3)$
0	.5	0	.5	0	.5
1	.5	1	.5	1	.5

c

Figure 16a/b/c

Figure 17 gives a diagrammatic representation of four possible structures, along with the probability distributions which these structures imply for the overall system.

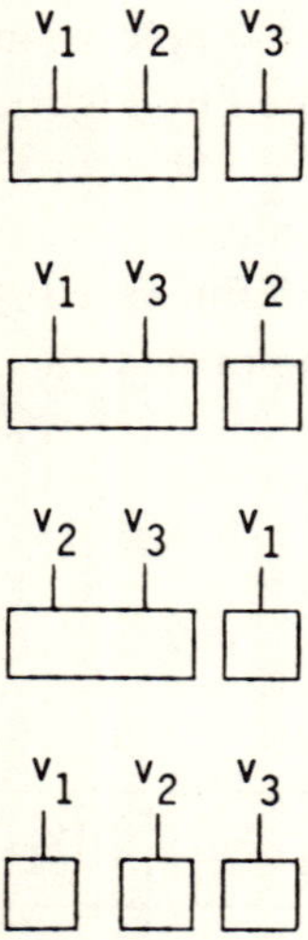

v_1	0	0	0	0	1	1	1	1
v_2 =	0	0	1	1	0	0	1	1
v_3	0	1	0	1	0	1	0	1
ST 1	.125	.125	.125	.125	.125	.125	.125	.125
ST 2	.125	.125	.125	.125	.125	.125	.125	.125
ST 3	.01	.24	.24	.01	.01	.24	.24	.01
ST 4	.125	.125	.125	.125	.125	.125	.125	.125

Figure 17

The structure ST 2, for example, essentially embodies the hypothesis that v_2 is independent of v_1 and v_3. Thus, the postulated probability of the sample 000, $p(000)$, is equal to $p(v_1v_3=00) \times p(v_2=0) = .25 \times .5 = .125$. To determine how reasonable each of the structures is, the corresponding row of Figure 17 is compared with the last column of Figure 1. It is easy to see that the overall behavior is perfectly reconstructable by considering v_2 and v_3 together as one subsystem, and v_1 as a separate subsystem, as in ST 3. It is also easy to see that none of the other three potential structures gives reasonably accurate reconstructability of the overall relation, reflecting the consideration made in Chapter 5 of the fact that all the 'constraint' of this system involves the interaction between variables v_2 and v_3 and thus the most acceptable decompositions cannot discard the information regarding their interaction. Constraint is used here in a more intuitive manner, and we reserve the technical use for the information theoretic formulations, though the existence of constraint among a set of variables is analogous to the divergence from probabilistic independence (Watanabe, 1969; Conant, 1976).

An alternate criterion to the difference in the probability distribution is based on direct consideration of constraint measured through the information theoretic concept of mutual information or transmission among a set of variables. In this case the divergence from independence among a set of variables is measured as the difference in entropy or uncertainty between that given by the joint distribution and that which would attain if the variables were independent (the sum of the entropies of the individual distributions). With respect to this criterion, better decompositions or structures are essentially those for which interaction among variables grouped together in subsystems is larger than the interaction between variables in different subsystems. In the three-variable system considered above, for example, any decomposition which put v_2 and v_3 in separate subsystems would naturally result in high transmission between these subsystems, indicating a poor structure system (cf. Broekstra, 1976, 1978).

The final criterion which we mention in this context is that of the increase in uncertainty regarding the next state of the overall system. This criterion was developed in detail in the preceding chapter.

It is important to realize that the simple example just considered was artificially constructed and is not very typical. That is, in general, it is unlikely that a set of system-defining variables will decompose perfectly into subsystems with respect to any meaningful criterion. In the case of the data on domestic conflict, and for most situations of interest in the social sciences, the cross-sectional nature of the data – even give attempts, such as with the transition to general image system, to control for systemic variation – further increases this unlikelihood.

Cognizance of this fact indicates that the most meaningful approach to structure determination from the perspective of general systems methodology is one which presents relevant information for sequences or sets of structure systems and allows the interpretation to be made in context of the investigation. An approach to this presentation is to consider structure determination as a process which starts with the overall relation represented as the complete binary relation on V. Elements are then removed from this relation, effecting a progressive refinement or disintegration of the overall relation. These elements are removed according to certain criteria whereby the process imposes a lattice structure on the set of structures considered (Klir, 1976; Klir and Uyttenhove, 1976; Cavallo and Klir, 1978b). The lattice may then be used to move through levels of increasing refinement proceeding at each level only to refinements of structures chosen as acceptable at the previous level and evaluating these structures at each level according to a chosen criterion and in light of whatever knowledge of the object that the investigator may possess. (See Klir and Uyttenhove, 1977, for some useful experimental guidelines; Klir, 1978b, for an application in ecological modelling.)

Another and complementary approach by which sets of structure systems could be presented for evaluation would involve specification and evaluation of all structures of a certain type (for example, disjoint subsets of the basic variables, i.e., the binary relation on V defining the structure would be an equivalence relation). Additionally, the best structures from each of a number of types could be evaluated. These approaches are especially important and meaningful in context of a specific investigation, as will become apparent through the description of their use in the rest of this section.

Requests of any type such as have been described are part of the definition of the set of requirements which constitute part of the systems problem. As we have mentioned, in cases where there is no suitably refined structure from which the overall relation may be perfectly reconstructed, the best policy toward structure determination – the most ideal sets of requirements – must be determined in context of the investigation.

Since the amount of material related to structure determination for the conflict system which we are describing is extensive we present here only the aspects which illustrate major points. In the last section we observed that obtaining the memoryless behavior served as a necessary part of identifiable problems associated with the determination of structure. Through use of the approach described in Klir and Uyttenhove (1976) on the memoryless behavior, a sequence of structure systems along with associated distances between structure-defined and actual overall probabilities is generated. While certain structure types which it makes sense to consider wind up not being evaluated in this scheme because of the nature of the lattice-imposing refinement process, other important information does emerge from the sequence.

The first structure system which emerges is that represented in Figure 18 (where it does not make a large difference we label attribute indicators with their symbols as given in Subsection 6.8.1.).

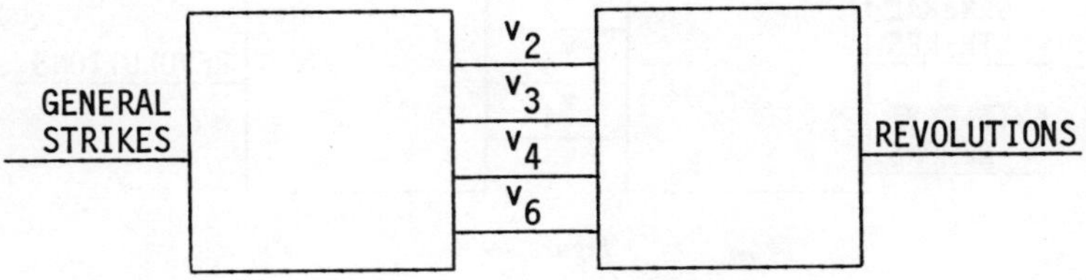

Figure 18

Because of the minimal amount of actual refinement the probability distribution distance is obviously small for this structure (less than 2 percent of that possible). In light of the extensive nature of the data, however – involving the full post-war period and using separate data points for each year – even the minimal statement represented by this structure system is not insignificant.

If we interpret the structure-derivation process in terms of dimensions as with previous studies, the representation here can be seen as describing the epistemological evolution (or breaking away) – within the constraints of our representation and choice of criteria and procedures – of the most significant subsystems from the parent overall system. This presents a complementary perspective, then, to that which confronts the overall system to determine '*the* structure' which generates or represents it. Figure 18 thus indicates that, starting from the conflict system as a whole, the most compelling direction of subsystem formation or separation tendency is that which distinguishes strike activity from that of revolutions. In context, it appears that the most distinguishing characteristic of these two dimensions centers around either the existence or the lack of major emphasis on the overthrow or replacement of the existing political regime. While it may appear intuitively obvious that this is the clearest differentiating mechanism for conflict activity, this could not have been determined from the bivariate correlations used, for example, by Rummel (1965). For the particular years of that study the correlation between general strikes and revolutions was not particularly low (.50) though, even if it were, there would be necessary qualifications due to the non-consideration of other attribute interactions.

The next level of refinement in the sequence is given in Figure 19. It is not unreasonable to interpret the break of Anti-Government Demonstrations from Revolutions as further supporting the tendency just described. We observe that all variables still interact through the triple: Government Crises, Purges, Riots.

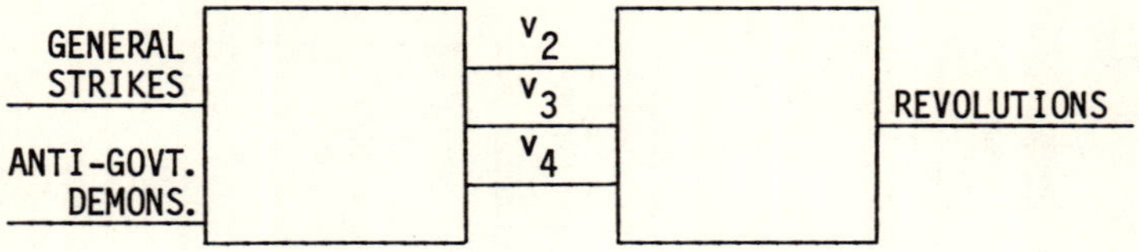

Figure 19

At the next level of refinement it is interesting to observe that, rather than maintain an emphasis toward the dual grouping we have described (as would be reasonable, say, with the split of Riots from the dimension which is oriented toward regime replacement) the Anti-Government Demonstrations exhibits the tendency – still in connection with Government Crises, Purges and Riots – toward separation also from General Strikes, as illustrated in Figure 20.

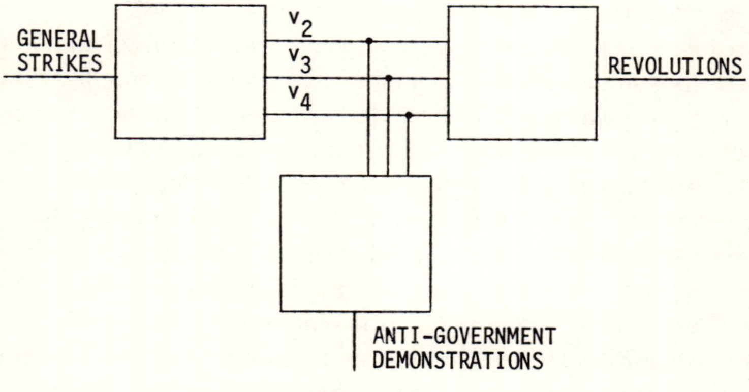

Figure 20

As the process continues, the General Strikes dimension disengages first from Government Crises, then Purges, and dissociates completely from all other dimensions after Riots separates from Revolutions. The resulting structure is given in Figure 21.

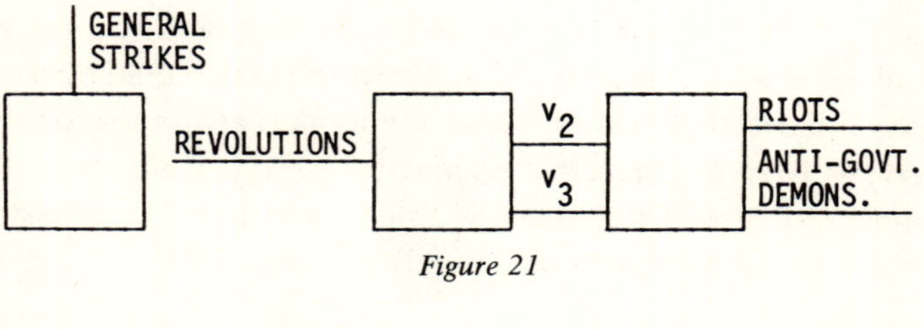

Figure 21

We observe that the separation of General Strikes at this level need not imply that the best structure of the form

$$\{v_i\},\ V - \{v_i\}$$

be that for which v_i = General Strikes (though in this example this is the case, with a negligible difference between that structure and that which only dissociates Purges). For interpretive significance we point out that, because of the nature of the next-structure determination process, individual variables do not split merely from another block of variables, but in general from a set of variables which itself constitutes a structure (or which *is* structured).

While there are perspectives from which this aspect is beneficial, it is also the case that questions relating to groups of attributes merely as an unstructured set may also be of interest. We again re-emphasize the intended flexibility of GSPS – of general systems research – in that such demands can easily be satisfied. For example, an important interpretation which may be given to the last structure illustrated is that Crises and Purge activity effectively serve as communication links through which structurally separated variables interact. Because of the procedure we have used in this structure determination, however, it is not possible to conclude that of all pairs of variables these two variables in fact best serve this communicative function. (We observe that with respect to this interpretation which emphasizes communicative aspects of the overall system it is of course possible to also give this interpretatiuon to structures determined through factor analysis. Strangely, there has been no concern with this in the past, the major emphasis being almost solely given to the separability of subsystems.)

Rather than address the problem of communicating subsystems by using only memoryless behavior, and in order to illustrate the benefits of GSPS's broad methodological perspective, we observe that in light of the interpretation it makes special sense to consider, in the context of structure evaluation, the effect of the past state of the overall system (or of the subsystems). In this regard we formulate the systems problem as a problem of the second kind involving comparison of pairs of (structure) systems and choose as criteria the overall next state entropy which accrues to various hypothesized structures. To simplify exposition we consider only structures of the form illustrated in Figure 22.

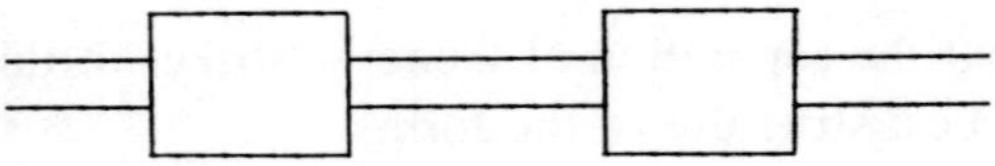

Figure 22

We are thus asking after the best decomposition which separates the overall system into two subsystems which interact or transmit information through two attributes. As criteria for comparison we use both of those which minimize the next state uncertainty which accrues to the overall system, the one which assumes that the next state of a subsystem is influenced by the current state of the overall system (which seems to be the most reasonable), and the other which considers the next state of a subsystem to be dependent only on the current state of that subsystem (the two criteria considered in detail in Chapter 5).

Results of the processing determined Government Crises and Riots as the two most suitable communication attributes with respect to each of the two criteria. In the cast of past state information transfer from the overall system, two structures were negligibly distant from each other, one of which was also determined best by the criterion which assumes that current states of a subsystem are affected only by past values of attributes associated with that subsystem. The best such structure is thus as illustrated in Figure 23.

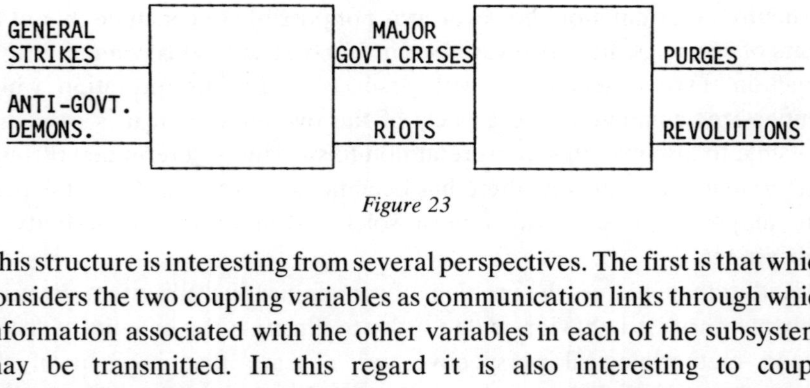

Figure 23

This structure is interesting from several perspectives. The first is that which considers the two coupling variables as communication links through which information associated with the other variables in each of the subsystems may be transmitted. In this regard it is also interesting to couple consideration of this structure with comments relating to the structure of Figure 18. That is, it is extremely reasonable to consider an overall conflict situation as exhibiting tendencies which are differentiated mainly by whether or not a change in political power (rather than satisfaction of some less extreme objectives) is the primary motivation. It is possible then to consider that Government Crises and Riots represent in some sense an intermediary stage or level of conflict activity between activities which are not overtly and primarily directed at elite replacement and those which are.

It is also interesting to consider this structure in light of the discussion of anomalous states in the section on memoryless behavior. Recall that the state involving Government Crises, Purges, Riots and Revolutions was the only state with significant variety of conflict activity which has a high frequency of appearance. The structure which has been determined here gives a very reasonable mode of explanation by which this anomaly appears not so anomalous. It is also significant to point out that none of the factor structures determined by past work and reported in Figures 12, 13 and 14 allow for the possibility of the significant interaction among these four dimensions. (The only one which comes close is Banks' factor structure for the combined data.)

Returning to the consideration of high-frequency states, recall that the only other such state expressing activity on even three dimensions is that involving Crises, Riots and Anti-Government Demonstrations. Obviously this state is also easily explainable in light of the structure of Figure 23, especially when considered in conjunction with the discussion following Figure 21. There, General Strikes was identified as exhibiting least interaction with other attributes, as being the one with the greatest tendency to 'separate.' Combined consideration of information resulting from two different problems related to structure systems thus generates an extremely reasonable context-related explanatory mechanism.

6.10 GSPS and Theory Construction

If the intention of this utilization of GSPS were to develop a 'theory' of conflict phenomena, it is clear that the considerations following Figure 23 offer an unassailable foundation from which to develop context-oriented 'theory construction.' In this sense they represent what Singer (1971) refers to as 'explanatory knowledge.' The derivation of this explanatory knowledge is thus roughly parallel to what we have referred to from the general systems perspective – and in light of the epistemological hierarchy of systems – as the determination of meaningful structure systems. It is also apparent that the explanatory knowledge so derived is (essentially by definition) in fact in keeping with the facts (Singer's 'existential knowledge' and our data system) and that it, in fact, 'predicts' or 'explains' what may be considered as anomalies, but what are perfectly reasonable, if not expected, in light of the potential 'theory.'

The intention of this chapter has, however, not been to construct a theory but to develop a well-organized, well-founded, process-oriented basis of inquiry and to demonstrate its applicability – thus demonstrating the power and utility of general systems methodology and of GSPS. Throughout this explication emphasis has been placed on the manner in which 'solutions' to particular problems invariably generate new systems problems when interpreted to context. This has in turn served to highlight two major epistemological premises of this book:

1. Value accrues to abstractions primarily through their interaction with specific systems. In this regard, we agree with Wittgenstein's (1956) recognition that value and truth are not static properties and that understanding is not a 'mental state,' but that value, truth and understanding result from process and use.
2. Truth and knowledge are as much subject to creation as to discovery.

6.11. Further Directions for Investigation

Given the essentially open-ended nature of inquiry it is reasonable, then, that the issues considered in the investigation of this chapter suggest further directions of useful study. Certain of these have been indicated in the text and others result from further consideration of material of this chapter in context of past studies. Of particular importance would be an extension or redefinition of the object system to include consideration of the ways in which conflict phenomena interact with other social and economic attributes and with more extensive forms of social and political repression (Firestone and Chadwick, 1972; Hibbs, 1973).

While such extensions will not be developed here it is useful to consider certain related questions in light of just the data system which is immediately accessible. The considerations give immediate further indication of the utility of GSPS.

Deutsch (1968) suggests that the most important functions of a nation are associated with its capability to change in response to events in its environment. Consideration of this fact in conjunction with conflict phenomena from a general systems and cybernetic perspective suggests an obvious interpretation of conflict indicators as communications or messages which specific overall politico-socio-economic systems transmit – mainly to the political regime in power. From one point of view such indicators can be seen as mechanisms for processes of national adaptation and growth.

In attempting to consider these effects, the different indicators obviously will not each have the same kind of significance. Past studies have attempted to cluster or classify countries according to different forms of violence. Rummel (1963), for example, uses the factor scores of each country on the dimensions he extracted and Banks (1972) runs Q-mode factor analyses. Rather than explore their findings in detail here, we merely point out that from the point of view described in the last paragraph there is an alternate or complementary perspective from which to consider the study of individual countries with respect to conflict activity.

In this regard we refer to the discussion in Section 5.4 regarding information processing limitations. If a country's response to conflict phenomena is an important consideration, then, in addition to considering the *quantity* of conflict or the characterization of the *major type* of conflict activity, the *variety* of that activity which responding bodies must face is likely to have a major effect on its ability to respond effectively.

Two conceptual approaches which have been discussed in Chapter 5 in the context of constraint provide immediate means for investigating this question. These are: the cylindrance decomposition of Ashby, and a slight adaptation of the measure of redundancy. As each of these concepts can be immediately made operational within GSPS, investigation from this perspective is trivial in context of the overall investigation.

The approach which has been followed in each case is to consider each country separately. For the cylindrance characterization the problem involves the determination of structure systems along with associated cardinality of the subsystems, given an initial behavior system consisting only of samples. In this case a low cylindrance indicates that – regardless of the overall quantity of violence to which the country must respond – there need not at any given time be a response which considers more forms of violence than the cylindrance. Thus, for example, a country with very high quantities of conflict on a particular dimension may be more able to efficiently adapt and positively respond than a country with lower quantities but for which these quantities are spread over various dimensions. Such a situation is indicated, for example, in comparing the situation of the United States with that of South Africa whose overall quantities of violence are roughly equivalent. For South Africa however the cylindrance is twice that of the United States (four vs. two), indicating a greater facility for the United States to turn its attention and resources to meeting its problems (e.g., witness the attention devoted and the response of the country to the purge activity of the McCarthy era).

In the case of the use of the redundancy measure, rather than considering variety along dimensions, the first characteristic measured is that of the entropy in the actual occurrence of states. Thus, a country with half of its occurrences indicating conflict activity, but where these occurrences are concentrated in two states, would be likely to be more capable of positive response to conflict than another with equal number of states indicating some activity, but where that activity is spread over many states.

For the comparison of individual countries a second and related measure was also used which expressed the actual entropy as a percentage of the maximum possible entropy given each country's number of states with conflict activity. This measure is thus analogous to the notion of redundancy as described in Chapter 5.

Comparisons of the rankings of the countries with respect to each of the criteria provide information through which more meaningful analyses of the effects of conflict on overall development can be made. For example, Belgium and the United Kingdom rank fourth and tenth respectively regarding the number of states exhibiting conflict activity but rank thirty-first and fortieth with respect to percentage of maximum entropy actually used. These countries thus can be expected to have had the opportunity to develop adequate means to respond to, or take advantage of, the indications represented by the conflict activity.

While significant political analysis regarding the emphases of this section requires more involved presentation and analysis of this data than would be appropriate here, the major intention has been satisfied. That has been basically to show that the concepts and procedures which constitute general systems research and GSPS provide a meaningful, operational, and useful mechanism for the symbiotic investigation of specific systems in conjunction with investigators whose expertise lies in the area of these specific systems.

Appendices

APPENDIX 1: Object-System Data (Raw Data) for Fifty-One Countries, 1946-1975

C10

0	0	0	0	0	0	0	0	0	0	0	0	0	0	0	0	0	0	0	0	0	0	0	0	0	0	0	0	0	0
1	0	0	1	0	0	0	0	0	0	0	0	0	0	0	0	0	0	0	0	0	0	0	0	1	0	0	0	0	0
0	0	0	0	0	0	0	0	0	0	0	0	0	0	0	0	0	0	0	0	0	0	0	0	0	0	0	0	0	0
0	0	0	0	0	0	0	0	0	1	0	0	0	0	0	0	0	0	0	0	0	0	0	0	0	0	0	0	0	0
0	0	0	0	0	0	0	0	0	0	0	0	0	0	0	0	0	0	0	0	0	0	0	0	0	0	0	1	0	0
0	0	0	0	0	0	0	0	0	0	0	0	0	0	0	0	0	0	0	0	0	0	0	0	1	0	0	0	0	0

C100

1	0	0	1	1	0	0	0	0	0	0	0	0	0	0	3	0	1	0	1	0	0	0	0	0	0	0	1	0	1
0	0	1	1	1	1	0	0	1	0	2	1	0	1	0	0	0	1	0	1	0	0	3	0	7	1	0	2	1	0
2	0	0	4	3	3	2	2	0	0	1	1	0	0	1	2	1	0	0	1	0	0	0	3	1	6	0	1	1	0
2	3	0	2	0	1	0	0	0	0	1	1	0	1	3	3	1	0	2	2	0	1	0	0	3	1	0	0	2	0
3	1	0	2	3	2	1	2	0	0	1	0	2	1	1	0	0	0	2	1	0	0	1	1	3	3	0	2	2	0
1	0	0	1	0	0	0	0	0	0	1	0	1	0	0	0	0	0	0	0	0	0	0	1	3	0	1	0	1	0

C1040

2	1	0	0	0	1	0	0	0	0	0	0	0	0	0	0	0	0	0	0	0	0	0	0	0	0	1	1	0	0
0	0	0	0	0	0	1	0	0	0	0	0	0	0	1	0	0	0	0	0	0	0	0	1	2	0	0	0	0	0
1	0	0	1	1	0	2	1	0	0	1	0	1	0	1	0	2	1	0	1	0	0	0	0	0	3	2	0	0	2
0	0	0	5	3	1	14	2	3	0	2	2	1	2	8	3	1	2	0	0	0	2	0	0	2	0	5	2	4	0
0	0	0	0	0	0	0	0	0	0	0	0	0	0	0	0	0	0	0	0	0	0	0	0	0	0	0	0	0	0
2	0	0	1	0	0	9	1	0	1	1	3	3	1	2	0	3	0	0	0	0	0	0	1	3	0	3	0	0	0

C1060

1	1	0	0	0	4	0	0	0	0	1	0	1	0	0	0	0	0	0	0	0	1	0	0	4	0	0	0	1	2
1	0	0	0	1	1	0	0	0	0	1	0	0	0	0	0	0	0	0	0	0	0	0	1	4	0	0	0	1	2
3	4	0	1	1	0	5	0	1	0	0	1	1	1	1	0	2	0	0	1	0	0	1	1	0	1	0	0	1	1
0	2	0	0	0	2	1	0	2	0	2	0	0	0	0	0	0	0	0	0	3	12	3	0	2	1	3	0	3	6
1	0	0	0	0	0	0	0	0	0	0	0	0	0	0	0	0	0	0	0	0	0	0	0	0	0	0	0	0	0
0	2	0	2	1	1	0	0	1	0	1	1	0	0	0	0	4	0	0	0	1	14	1	4	8	0	1	1	1	9

C1091

0	0	0	1	0	0	0	0	0	0	0	0	0	0	0	0	0	0	0	0	0	0	0	0	1	0	0	0	0	0
0	0	0	0	0	0	0	0	0	0	0	0	0	0	0	0	0	2	0	0	0	0	0	0	0	1	1	0	0	0
0	0	0	0	0	0	0	0	0	0	0	0	0	0	0	0	0	0	0	0	0	0	0	0	0	0	0	0	0	0
0	0	0	0	0	0	0	0	0	0	0	0	0	0	0	0	0	0	0	0	0	0	0	0	0	0	0	0	0	0
0	0	0	0	0	0	0	0	0	0	0	0	0	0	0	0	0	0	0	0	0	0	0	0	0	0	0	0	0	0
0	0	0	0	0	1	0	1	0	0	0	0	0	0	0	0	0	0	0	0	0	0	0	0	0	0	0	0	0	0

C1092

0	0	0	0	0	0	0	0	0	0	0	0	0	0	0	0	0	0	0	0	0	0	0	0	0	1	0	0	0	0
0	0	0	0	0	0	0	0	0	0	0	1	0	1	0	0	0	0	0	0	0	0	0	0	1	0	0	0	0	0
0	0	0	0	0	0	0	0	0	0	0	0	0	1	0	0	0	0	0	0	0	0	0	0	0	0	0	0	0	0
0	0	0	0	0	7	0	0	0	0	0	0	0	0	0	0	0	0	0	0	0	1	1	0	0	0	0	0	0	0
0	0	0	0	0	0	0	0	0	0	0	0	0	0	0	0	0	0	0	0	0	0	0	0	0	0	0	0	0	0
0	0	0	0	0	0	0	0	0	0	0	0	0	0	0	0	0	0	0	0	0	0	0	0	0	0	1	0	0	0

C1100
```
0 0 0 0 0 0 0 0 0 0 0 0 0 0 0 0 0 0 0 0 0 0 0 0 0 0 0 0 0 0
0 0 0 0 0 0 0 0 0 0 0 0 0 0 0 0 0 0 0 0 0 0 0 0 0 0 0 0 0 0
0 0 0 0 0 0 0 1 0 0 0 0 0 0 0 0 0 0 0 0 0 0 0 0 0 0 0 0 0 0
0 0 0 0 0 0 0 0 0 0 0 0 0 0 0 0 0 0 0 0 0 0 0 0 0 0 0 0 0 0
0 0 0 0 0 0 0 0 0 0 0 0 0 0 0 0 0 0 0 0 0 0 0 0 0 0 0 0 0 0
0 0 0 0 0 0 0 0 0 0 0 0 0 0 0 0 0 0 0 0 0 0 0 0 0 0 0 0 0 0
```

C1130
```
0 0 0 0 0 0 0 0 0 0 0 0 0 0 0 0 0 0 0 0 0 0 0 0 0 0 1 0 0 1
1 0 0 1 0 0 0 0 0 0 0 1 0 0 0 0 0 0 0 0 0 0 1 0 2 1 0 1 1 1
1 0 1 1 1 2 0 1 1 0 0 0 2 0 0 1 0 1 0 0 0 0 0 0 0 0 0 1 0 0
0 0 0 1 0 0 0 0 0 0 0 0 0 0 0 0 0 0 0 0 0 0 0 0 0 0 0 1 2 2
0 2 0 1 1 2 1 0 0 0 0 1 1 0 0 0 0 0 0 0 0 0 0 0 0 1 0 1 0 0
0 0 0 0 0 0 0 0 0 0 0 0 0 0 0 0 0 0 0 0 0 0 0 0 1 0 0 1 3 2
```

C1170
```
0 0 0 0 0 0 0 0 0 0 0 0 0 0 0 0 0 0 0 0 0 0 0 0 0 0 0 0 0 0
0 1 1 2 0 1 0 0 0 0 1 0 1 0 1 0 1 2 0 1 0 0 0 0 2 2 2 2 1 0
1 0 0 0 0 0 1 0 0 1 0 0 1 0 0 2 2 2 0 0 0 0 0 0 0 0 0 2 0 0
1 0 0 0 0 0 0 0 0 1 0 2 0 0 4 0 1 1 0 0 0 3 1 4 4 1 0 0 0 5
0 0 0 0 0 0 0 0 0 0 0 0 0 0 1 0 1 1 0 0 0 0 0 0 0 0 0 0 0 0
0 1 1 0 0 0 0 0 0 0 0 0 0 0 6 0 0 2 0 0 0 0 1 0 1 0 0 0 0 4
```

C1190
```
0 0 0 0 0 0 0 0 0 0 0 0 0 0 0 0 0 0 0 0 0 0 0 0 0 0 0 0 0 0
0 0 0 0 0 0 0 0 0 0 0 0 0 0 0 0 1 0 0 0 0 0 0 0 1 0 0 0 0 0
0 0 0 1 1 2 1 4 1 2 0 2 0 0 1 0 1 1 2 1 0 0 1 1 0 0 3 3 1 0
0 0 0 0 0 1 0 0 0 0 1 1 0 0 0 1 1 0 0 0 0 0 0 0 0 0 1 0 0 0
0 0 0 0 1 0 0 0 0 0 0 0 0 0 0 0 0 0 0 0 0 0 0 0 0 0 0 0 0 0
0 0 0 0 0 0 0 0 0 0 1 0 0 0 0 0 1 1 0 0 0 0 0 0 1 0 0 0 3 0
```

C1 20
```
1 0 0 1 0 0 0 0 2 0 0 0 0 0 0 0 1 1 0 0 0 0 0 0 0 0 0 0 0 0
1 0 1 0 0 1 0 1 3 2 1 0 0 0 0 1 2 2 0 0 0 0 1 1 3 0 0 0 0 0
1 0 1 0 2 0 4 0 1 2 2 2 1 0 0 0 0 0 2 0 0 0 1 34 1 9 1 1 1 0
0 1 2 2 0 2 6 1 3 0 0 0 6 4 2 2 1 1 3 0 1 0 2 1 0 1 0 0 0 0
0 1 0 0 0 2 0 0 1 1 1 0 0 1 0 0 0 1 1 0 0 0 0 0 0 0 0 0 0 0
0 0 0 0 0 0 1 0 1 0 0 0 1 0 0 1 0 0 0 1 1 1 0 1 2 0 0 0 2 0
```

C1 210
```
0 1 0 0 0 0 0 0 0 0 0 0 0 0 0 0 0 0 0 0 0 0 0 0 3 1 1 2 1 0
0 2 0 1 0 1 0 0 0 0 2 0 0 0 0 0 0 3 0 0 0 0 0 0 2 0 0 0 0 0
0 0 0 0 0 0 0 0 0 0 0 0 0 0 0 0 0 0 0 0 0 0 0 0 0 0 0 0 0 0
1 3 0 2 2 1 1 1 1 1 2 1 2 0 0 0 2 6 0 0 0 0 0 2 0 0 0 0 0 0
0 0 0 0 0 0 0 0 0 0 0 0 0 0 0 0 0 0 0 0 0 0 1 0 1 0 0 0 0 0
1 0 0 0 0 1 0 1 0 1 0 0 0 1 6 2 2 5 4 1 0 0 6 0 2 0 1 0 4 0
```

C1220

2	1	0	0	0	0	0	0	0	0	0	0	0	0	0	0	0	0	0	0	0	0	0	0	0	0	0	0	0	0
0	0	0	0	0	0	1	0	0	0	0	0	0	0	0	0	0	0	0	0	0	0	0	0	0	0	0	1	0	0
0	0	1	1	1	1	1	0	1	0	0	0	0	0	0	0	0	0	0	0	0	0	0	0	0	0	0	0	0	0
6	0	3	2	0	1	0	1	0	0	4	4	3	2	8	7	4	19	14	12	16	55	48	4	5	1	0	0	0	0
0	0	0	0	0	0	0	0	0	0	0	0	0	0	0	0	0	0	0	0	0	0	0	0	0	0	0	0	0	0
0	2	1	3	0	3	1	2	1	2	4	2	1	2	22	3	7	60	11	13	13	28	37	30	20	2	2	3	0	8

C1240

0	0	0	0	0	2	0	0	0	0	0	0	0	0	1	0	0	0	0	0	0	1	2	0	0	0	1	1	0	0
0	0	1	1	0	0	0	0	0	0	1	0	0	0	0	0	0	0	0	1	0	1	2	1	2	0	1	2	1	0
0	0	0	0	0	0	0	0	0	0	0	1	0	0	0	1	0	0	1	0	0	0	0	1	0	0	0	0	1	0
1	0	0	0	0	0	0	0	0	0	0	0	0	0	0	3	0	0	2	0	0	0	5	2	1	1	0	1	0	0
1	0	0	0	0	0	0	0	0	0	0	0	0	0	0	0	0	0	0	0	0	0	0	0	0	0	0	0	0	0
0	0	0	0	0	0	0	0	0	0	0	0	2	0	1	0	1	0	0	0	0	0	3	0	1	0	0	2	0	0

C1250

0	0	0	3	1	0	0	0	0	0	0	0	1	0	0	0	0	0	0	0	0	0	0	0	0	0	0	0	0	0
0	0	2	1	0	0	0	0	0	0	0	0	1	0	2	0	0	1	1	0	0	0	0	0	1	0	0	0	0	0
1	1	1	1	1	0	3	1	1	0	0	1	4	0	5	2	1	0	0	0	0	0	0	0	0	0	0	0	0	0
3	0	0	0	0	1	0	0	0	0	0	0	6	0	12	0	1	0	0	0	3	0	0	1	1	1	0	0	0	0
4	2	1	0	0	1	1	0	0	0	0	0	4	0	3	2	2	0	0	0	1	0	0	0	0	0	0	0	0	0
0	0	0	0	0	0	0	0	0	0	0	1	1	0	2	3	0	0	0	0	0	0	0	1	0	0	0	0	0	0

C1290

1	0	0	0	0	0	0	0	0	0	0	0	0	0	0	0	0	0	0	0	0	0	0	0	0	0	0	0	0	0
0	0	0	0	0	0	0	0	0	0	0	0	0	0	0	0	0	0	0	0	0	0	0	0	0	1	0	0	0	0
2	1	2	5	1	2	4	0	1	0	0	0	2	0	0	0	0	0	0	0	1	2	1	0	0	1	6	0	1	3
1	0	0	0	0	0	0	2	0	0	0	0	0	0	0	1	0	0	0	0	0	0	2	0	0	1	0	0	0	0
0	0	0	0	0	0	0	0	0	0	0	0	0	0	0	0	0	0	0	0	0	0	0	0	0	0	0	0	0	0
0	0	0	0	0	0	0	0	1	0	0	0	0	0	0	0	0	0	0	0	0	0	0	1	0	0	0	0	0	0

C130

0	0	0	0	0	0	0	0	0	0	0	0	0	0	0	0	0	0	0	0	0	0	0	0	0	0	0	0	0	0
1	0	0	0	0	1	0	0	0	0	0	0	0	0	0	0	0	0	0	0	0	0	0	0	0	0	0	0	0	0
2	4	1	6	1	1	1	0	0	0	1	3	0	0	0	0	1	0	0	1	0	0	0	1	0	0	0	0	0	0
0	0	0	0	0	0	0	0	0	0	0	0	0	0	0	0	0	1	1	0	0	0	0	0	0	0	0	0	0	0
0	2	0	0	0	0	0	0	0	0	0	0	0	0	0	0	0	0	0	0	0	0	0	0	0	0	0	0	0	0
0	0	0	0	0	0	0	0	0	0	0	0	0	0	0	0	0	0	0	0	0	0	0	0	0	0	0	0	0	0

C180

0	0	0	0	0	0	0	0	0	0	0	0	0	0	0	0	0	0	0	0	0	0	0	0	1	0	0	0	0	1
0	0	0	0	0	0	0	0	0	0	0	0	1	0	0	0	0	2	0	0	0	0	0	1	5	0	0	0	0	0
0	0	0	0	0	0	0	0	0	0	0	0	0	0	0	0	0	0	0	0	0	0	0	0	0	1	0	0	0	0
1	0	1	2	0	0	0	0	0	1	0	0	0	0	0	0	0	1	1	1	0	0	0	1	0	0	0	0	0	0
0	0	0	0	0	0	0	0	0	0	0	0	0	0	0	0	0	0	0	0	0	0	0	0	0	0	0	0	0	0
0	0	0	0	0	1	0	0	0	0	0	0	1	0	0	0	0	0	0	0	1	0	0	1	1	1	0	0	0	0

*C*20

0	0	0	0	0	0	0	0	0	0	0	0	0	0	0	0	0	0	0	0	0	0	0	0	0	0	0	0	0	0
0	0	0	1	0	1	0	1	0	0	0	0	0	0	0	0	0	0	0	0	0	0	0	0	0	0	0	0	0	0
1	1	2	1	1	2	2	0	0	0	0	0	0	0	0	0	0	0	0	0	0	0	0	0	0	0	0	0	0	0
0	0	0	0	0	0	0	0	0	0	0	0	0	0	0	0	0	0	0	0	0	0	0	0	0	0	0	0	0	0
0	0	0	0	0	1	0	0	0	0	0	0	0	0	0	0	0	0	0	0	0	0	0	0	0	0	0	0	0	0
0	0	0	0	0	0	0	0	0	0	0	0	0	0	0	0	0	0	0	0	0	0	0	0	0	0	0	0	0	0

*C*220

3	1	0	0	1	0	0	0	1	1	1	0	0	0	0	0	0	0	0	1	0	1	0	1	0	0	1	2	0	0
0	1	0	1	0	2	0	0	1	2	0	0	0	0	0	0	0	1	0	0	0	0	1	1	1	3	3	1	0	0
0	2	2	2	0	0	0	0	2	1	1	0	0	0	0	1	0	0	0	0	0	0	0	0	0	1	0	3	2	0
1	0	0	2	0	0	2	0	0	0	0	2	1	0	2	1	0	1	0	2	2	1	0	3	2	1	7	2	0	0
0	0	1	1	0	0	0	0	0	0	0	0	0	0	0	0	0	0	0	0	0	0	0	1	1	0	0	1	0	0
0	0	0	0	0	0	0	0	0	0	0	0	1	0	1	0	0	0	0	1	1	0	0	0	1	1	6	3	1	0

*C*230

0	0	0	0	0	0	0	0	0	0	0	0	0	0	0	0	0	0	0	0	0	1	0	0	0	0	0	0	0	0
2	0	0	0	0	0	0	0	0	0	0	0	0	0	0	0	2	0	0	0	0	1	0	1	1	1	0	0	1	0
0	2	0	0	1	5	5	0	4	3	0	1	4	1	0	0	0	0	0	0	2	1	0	3	0	1	0	1	3	0
3	0	2	0	2	0	1	1	1	0	0	0	0	0	2	0	0	0	0	0	0	32	8	0	1	0	0	0	0	2
2	1	1	3	1	1	1	1	0	0	1	1	0	1	0	0	0	0	0	0	0	0	0	0	0	0	0	0	0	0
0	0	1	0	0	0	0	0	0	0	0	1	0	0	1	0	0	0	0	0	0	7	0	0	2	0	0	0	0	1

*C*240

0	0	0	0	0	0	0	0	0	0	0	0	0	0	0	0	0	0	0	0	0	0	0	0	0	1	0	0	0	0
1	0	0	2	0	0	1	1	0	0	1	1	0	0	0	0	1	2	0	1	0	0	1	0	4	2	0	0	0	0
0	0	0	1	0	0	0	0	1	0	0	2	2	0	0	0	0	1	0	0	0	0	0	0	0	0	0	0	0	0
2	3	3	5	0	0	1	0	4	0	0	4	1	0	0	0	0	1	1	1	1	2	0	2	1	2	0	0	0	1
0	0	1	0	1	0	0	1	0	0	1	1	1	0	0	0	0	0	0	0	0	0	0	0	0	0	0	0	0	0
0	0	0	0	0	0	1	0	0	1	0	0	0	0	0	0	0	1	0	0	0	0	0	0	1	0	0	0	0	1

*C*270

0	2	0	0	0	0	0	0	0	0	0	0	0	0	0	0	0	0	0	0	0	0	0	0	0	0	0	0	0	0
1	0	0	1	0	0	1	0	0	0	0	0	0	0	0	0	0	0	0	0	0	0	0	0	0	0	0	0	0	0
0	0	0	0	1	0	0	0	0	1	0	0	0	0	0	0	0	0	0	0	0	0	0	0	0	0	0	0	0	0
0	3	1	0	2	0	0	0	0	0	0	0	0	0	0	0	0	0	0	0	0	0	0	1	0	0	0	0	0	0
1	0	3	2	0	1	0	0	1	1	0	0	0	0	0	0	0	0	0	0	0	0	0	0	0	0	0	0	0	0
0	1	0	0	0	0	0	0	0	0	0	0	0	0	0	1	0	0	0	0	0	0	0	0	0	0	0	0	0	0

*C*280

0	4	1	0	0	0	0	0	0	0	0	1	0	1	0	0	0	0	0	0	0	0	0	0	0	0	0	0	0	0
0	0	0	1	0	0	1	0	0	0	1	0	0	2	0	0	0	0	0	0	0	0	0	0	1	0	0	0	0	0
1	4	0	2	0	0	4	0	3	0	2	2	4	5	5	3	0	0	0	1	1	0	0	1	0	2	0	0	0	0
3	1	2	3	0	1	0	2	0	1	0	0	2	2	4	2	1	0	0	0	1	0	0	0	0	0	0	0	0	0
1	0	0	0	0	0	2	1	0	0	2	3	1	2	1	1	1	0	0	0	0	0	0	0	0	0	0	0	0	0
0	0	1	0	0	0	2	0	0	0	0	2	1	0	3	2	2	0	0	0	0	0	0	0	0	0	0	0	0	0

*C*320

1	0	0	0	0	0	0	0	0	0	0	0	0	0	0	0	0	0	0	0	0	0	0	0	0	0	0	0	1	0
0	0	0	0	0	0	0	1	0	0	0	1	0	0	0	0	0	0	0	0	1	1	0	0	0	0	0	2	0	0
0	0	0	0	0	0	0	0	0	0	0	0	0	0	0	0	0	0	0	0	0	0	0	0	0	0	0	0	0	0
0	0	0	0	0	1	0	0	0	0	0	0	0	0	0	0	0	0	0	0	0	0	0	0	0	0	0	0	0	0
0	0	0	0	0	0	0	0	0	0	0	0	0	0	0	0	0	0	0	0	0	0	0	0	0	0	0	0	0	0
0	0	0	0	0	0	0	1	0	0	0	0	0	0	0	0	0	0	0	1	0	0	1	0	0	0	1	0	0	0

*C*340

0	0	0	0	0	0	0	0	0	0	0	0	0	1	0	0	0	0	0	0	0	0	0	0	0	0	0	0	0	0
1	0	0	1	0	1	1	0	0	0	0	0	0	1	0	1	1	0	0	1	0	0	0	0	2	0	0	0	0	0
1	0	1	1	0	0	1	0	1	2	0	0	0	0	1	1	1	0	1	0	1	0	0	0	0	0	0	0	0	0
0	0	1	0	0	0	2	2	0	0	0	0	1	2	1	2	1	0	0	1	2	1	0	2	1	0	0	0	0	0
2	6	0	1	0	0	1	0	1	0	0	0	0	0	0	1	1	1	0	0	1	0	0	0	1	1	1	0	0	1
0	0	0	0	0	0	0	0	0	0	0	0	0	0	0	0	0	0	2	2	2	0	0	0	1	0	0	0	0	0

*C*350

1	0	0	0	0	0	0	0	0	0	0	0	0	0	0	0	0	0	0	0	0	0	0	0	0	0	0	0	0	0
0	0	0	0	0	1	0	0	0	0	0	0	0	0	0	0	0	0	0	0	0	0	0	0	0	0	0	0	0	0
1	0	0	0	0	0	0	0	0	0	0	0	0	0	2	0	1	0	0	0	0	0	0	0	0	0	0	0	0	0
1	0	0	1	0	0	0	0	0	0	0	0	0	0	0	0	0	0	0	0	1	0	0	0	0	0	0	0	0	0
0	1	1	0	0	0	1	0	0	0	0	0	0	0	1	1	0	0	0	0	0	0	0	0	0	0	1	0	0	0
0	0	0	0	0	0	0	0	0	0	0	0	0	0	0	0	0	0	0	0	1	0	0	0	0	0	1	0	0	0

*C*380

0	2	2	0	0	0	0	0	0	0	0	0	0	0	0	0	0	0	0	0	0	0	0	0	0	0	0	0	0	0
1	0	0	1	3	0	2	2	2	0	1	2	3	0	1	1	0	2	1	0	0	0	0	0	1	1	2	0	0	1
2	0	0	1	0	1	0	0	0	0	0	0	0	0	0	0	0	1	0	0	0	0	0	0	0	0	0	0	0	0
0	0	2	1	0	0	0	0	0	0	0	0	0	0	0	0	1	0	0	0	0	0	0	0	0	0	0	0	0	0
0	0	0	0	0	0	0	0	0	0	0	0	0	0	0	0	0	0	0	0	0	0	0	0	0	0	0	0	0	0
0	0	0	0	0	0	0	0	0	0	0	0	0	0	0	0	0	0	0	0	0	0	0	0	0	0	0	0	0	0

*C*390

0	5	2	1	4	2	2	2	0	1	0	1	0	0	0	2	1	1	0	0	2	1	1	2	0	1	0	2	3	1
2	5	2	2	3	5	3	4	1	3	1	3	1	1	1	0	2	0	0	0	0	0	1	0	1	1	0	0	0	0
0	0	0	0	1	0	2	1	0	0	0	0	1	1	2	1	4	0	0	0	0	0	1	0	0	0	0	0	0	0
1	8	11	3	3	3	3	3	0	2	5	1	3	0	5	3	1	4	0	0	3	2	12	1	6	2	0	1	4	3
0	1	0	1	0	0	0	0	0	0	0	0	0	0	0	0	0	0	0	0	0	0	0	0	0	0	0	0	0	0
0	7	0	4	2	1	1	1	1	1	0	0	1	0	4	1	4	5	0	0	1	1	5	1	4	1	0	1	2	3

*C*40

1	0	0	2	0	2	0	0	0	1	2	2	1	6	0	3	1	0	0	0	0	1	0	13	7	1	0	1	1	2
0	0	0	0	1	1	0	0	0	1	2	1	1	5	2	0	6	3	1	1	0	1	1	1	4	0	1	0	1	2
2	0	0	1	1	2	3	1	3	5	4	2	4	5	1	1	4	2	0	1	1	0	1	4	0	1	0	1	1	0
6	0	0	1	0	0	0	2	4	5	2	1	2	2	1	1	2	0	1	0	2	0	1	15	1	1	0	3	8	3
0	0	0	0	0	0	1	0	0	3	0	0	0	1	2	1	5	1	0	0	1	0	0	0	2	2	0	0	1	1
0	0	0	0	0	0	0	0	1	5	0	1	4	0	0	0	0	0	0	1	2	0	2	0	2	0	5	2	3	1

C450

1	0	0	0	0	0	0	1	0	0	0	0	0	0	0	0	0	0	0	0	0	0	0	0	0	0	0	0	1	0
0	1	0	2	2	2	1	0	1	0	0	0	1	0	0	1	0	2	0	3	2	1	0	0	0	0	0	1	1	0
1	1	0	0	0	1	3	0	0	0	0	0	0	0	0	0	0	0	0	0	0	2	5	1	1	5	2	3	1	3
6	0	1	0	0	0	1	1	1	0	0	0	0	0	0	0	1	1	0	5	2	2	0	0	0	0	0	0	1	0
0	1	1	1	0	0	0	0	0	0	0	0	0	0	0	0	0	0	0	0	0	2	0	0	0	0	0	1	0	1
0	0	0	0	1	0	1	0	0	0	0	0	0	0	0	0	0	2	0	0	2	0	2	0	0	1	1	4	1	1

C460

0	0	0	0	1	0	0	0	0	0	0	0	0	0	0	0	0	0	0	0	0	0	0	0	0	0	0	0	0	0
0	0	0	2	0	2	0	0	1	0	1	0	0	0	0	0	1	2	0	0	0	1	0	0	7	0	0	0	1	0
0	0	0	1	0	0	13	1	6	2	3	0	0	0	0	2	1	1	0	0	1	0	1	0	0	0	0	0	0	0
1	1	0	0	1	2	2	3	4	1	1	1	0	0	2	0	0	0	0	1	0	0	0	0	1	0	0	0	1	0
1	0	1	2	0	0	0	1	1	1	0	1	0	0	2	0	1	1	0	0	0	0	0	0	1	0	0	0	0	0
1	0	0	1	0	1	8	0	2	1	2	0	0	0	3	0	0	0	0	0	0	0	0	0	0	1	0	0	4	0

C490

0	0	0	0	0	0	0	0	1	0	1	4	0	0	0	0	0	0	0	0	0	0	0	0	0	0	0	0	0	0
1	0	0	2	0	0	0	0	0	0	1	5	0	2	0	0	0	1	0	0	0	0	0	0	0	1	0	0	0	0
0	0	0	1	0	0	0	0	1	0	1	1	3	0	2	0	0	1	1	0	0	0	1	0	1	0	0	0	0	0
3	0	0	0	0	0	0	0	0	0	0	5	0	0	0	1	0	2	0	0	0	0	0	0	1	0	0	0	0	0
1	0	0	0	0	0	0	0	0	0	0	2	1	0	0	0	0	1	0	0	0	0	0	0	1	0	0	0	0	0
1	0	0	1	0	0	0	0	0	0	0	1	0	1	0	1	0	0	0	0	0	0	0	0	0	0	0	0	0	0

C50

0	0	0	1	0	1	0	0	0	0	0	0	0	0	0	0	0	0	0	0	0	0	0	0	0	0	0	0	0	0
0	0	0	0	0	0	0	0	0	0	0	0	0	0	0	0	0	0	0	0	0	0	0	2	0	0	0	0	1	2
0	0	0	0	0	0	0	0	0	0	0	0	0	0	0	0	0	0	0	0	0	0	0	0	0	0	0	0	0	0
0	0	1	1	0	0	0	0	0	0	0	1	0	0	1	0	0	0	0	0	0	0	0	1	0	0	0	0	0	0
0	0	0	0	0	0	0	0	0	0	0	0	0	0	0	0	0	0	0	0	0	0	0	0	0	0	0	0	0	0
0	0	0	0	0	0	0	0	0	0	0	0	0	0	0	0	0	0	0	0	0	0	0	2	5	0	0	0	0	0

C500

0	0	0	0	0	0	0	0	0	0	0	0	0	0	0	0	0	0	0	0	0	0	0	0	0	0	0	0	0	0
0	0	0	0	0	0	0	1	0	0	0	0	1	0	0	0	0	0	0	0	0	0	0	0	1	0	0	0	0	0
0	0	0	0	0	0	0	1	0	0	0	0	0	1	0	0	0	0	0	0	0	0	0	0	0	0	0	0	0	0
0	0	1	0	0	0	0	1	0	0	1	0	0	0	0	0	0	0	0	0	0	0	0	1	0	0	0	0	0	0
0	1	0	0	0	0	0	0	0	2	1	0	2	0	0	0	1	0	0	0	0	0	0	0	0	0	1	0	0	1
0	0	0	0	0	0	0	0	0	0	0	0	0	0	0	0	0	0	0	0	0	0	0	1	0	0	0	0	0	1

C540

1	0	0	0	0	1	0	0	0	0	0	0	0	0	0	0	0	0	0	0	0	0	0	0	0	0	0	0	0	0
2	0	2	2	1	4	1	4	0	0	0	0	0	1	1	1	0	1	0	0	0	0	0	0	0	0	0	0	0	0
1	0	1	2	1	1	2	2	3	0	1	0	1	1	1	2	0	1	0	0	0	0	0	0	0	1	0	0	0	0
3	1	1	0	0	7	8	13	4	0	0	1	0	0	1	2	1	3	0	0	3	0	0	0	0	0	0	0	0	1
1	0	0	1	0	0	1	3	0	0	0	0	0	0	0	0	0	1	0	0	0	0	0	0	0	0	0	0	0	0
0	0	1	1	0	3	4	3	0	0	0	0	0	0	1	4	1	1	0	0	1	0	0	2	0	0	0	0	0	2

*C*570

3	7	2	5	6	1	2	2	4	0	0	0	0	0	0	0	0	0	0	0	0	0	1	2	6	0	2	0	6	0
0	3	0	0	1	1	0	2	2	2	0	2	1	1	3	0	1	4	2	0	1	0	3	0	5	0	2	2	2	1
0	1	0	1	0	1	1	0	0	0	0	0	0	0	0	0	0	0	0	0	0	0	0	0	0	2	0	0	0	0
5	19	9	11	9	1	4	2	3	0	1	1	0	3	8	1	4	0	0	0	2	1	7	4	6	10	6	0	2	8
1	2	0	0	2	0	0	0	0	0	0	0	0	0	0	0	0	0	0	0	0	0	0	0	0	0	0	0	0	0
4	9	0	2	1	3	6	3	1	1	1	1	1	0	0	0	0	2	0	0	1	0	4	0	3	6	3	0	2	4

*C*62

0	0	0	0	0	0	0	0	0	0	4	0	0	0	0	0	0	0	0	0	0	0	0	0	0	0	0	0	0	0
0	1	0	0	0	0	0	1	0	0	4	1	0	0	0	0	0	0	0	0	0	0	0	0	0	0	0	0	0	0
0	2	2	3	2	4	3	1	0	2	2	2	1	0	0	0	2	0	0	1	0	0	0	0	0	0	0	0	0	0
0	2	1	0	0	0	0	0	0	0	2	1	0	0	0	0	0	0	0	0	0	0	0	0	0	0	0	0	0	0
1	1	0	0	0	0	0	0	0	0	1	0	0	0	0	0	0	0	0	0	0	0	0	0	0	0	0	0	0	0
0	0	0	0	1	0	0	0	0	0	3	0	0	0	0	0	0	0	0	0	0	0	0	0	0	0	1	0	0	0

*C*690

0	0	0	0	0	0	0	0	0	0	0	0	0	0	0	0	0	0	0	0	0	0	0	0	0	0	0	0	0	0
0	0	0	0	0	0	0	0	0	0	0	0	0	0	0	0	0	0	0	0	0	0	0	0	0	0	0	0	0	0
0	0	0	0	0	0	0	0	0	0	0	0	0	0	0	1	0	0	0	0	0	0	0	0	0	0	0	0	0	0
0	0	0	0	0	0	0	0	0	0	0	0	0	0	0	1	0	1	0	0	0	0	0	0	0	0	0	0	0	0
0	0	0	0	0	0	0	0	0	0	0	0	0	0	0	0	0	0	0	0	0	0	0	0	0	0	0	0	0	0
0	0	0	0	0	0	0	0	0	0	0	0	0	0	0	0	0	0	0	0	0	0	0	0	0	0	0	0	0	0

*C*720

0	0	0	0	0	0	0	0	0	0	0	0	0	0	0	0	0	0	0	0	0	0	0	0	0	0	0	0	0	0
0	1	0	0	0	0	0	0	0	0	0	0	1	0	0	0	0	0	0	0	0	0	0	0	0	0	0	0	0	0
1	0	0	0	0	0	0	0	0	0	0	0	0	0	0	0	0	0	0	0	0	0	0	0	0	0	0	0	0	0
0	0	0	0	0	0	0	0	0	0	0	0	0	0	0	0	0	0	0	0	0	0	0	0	0	0	0	0	0	0
0	0	0	0	0	0	0	0	0	0	0	0	0	0	0	0	0	0	0	0	0	0	0	0	0	0	0	0	0	0
0	0	0	0	0	0	0	0	0	0	0	0	0	0	0	0	0	0	0	0	0	0	0	0	0	0	0	0	0	0

*C*80

0	1	1	0	1	0	0	0	1	1	0	0	0	0	2	2	0	0	1	0	0	0	0	0	0	0	1	0	0	0
2	0	2	0	2	0	1	0	0	0	0	0	1	0	0	1	0	1	0	1	1	0	1	0	1	0	2	0	1	0
0	0	0	0	0	0	0	0	1	0	0	0	0	0	0	1	0	0	0	0	0	0	0	0	0	0	0	0	0	0
0	1	0	0	4	0	0	1	0	5	0	0	0	0	3	3	1	1	0	1	1	0	2	0	0	0	0	0	0	0
0	0	0	0	0	0	0	0	0	0	0	0	0	0	0	0	0	0	0	0	0	0	0	0	0	0	0	0	0	0
0	0	0	1	0	1	1	0	0	3	0	0	1	0	1	0	0	0	0	0	0	1	0	1	0	0	0	0	0	0

*C*810

1	0	0	0	0	0	1	0	0	0	0	0	1	0	0	0	0	0	0	0	0	0	1	0	0	0	0	0	0	0
0	0	0	0	0	0	0	0	0	0	0	0	0	0	0	0	0	0	0	0	0	0	1	0	0	0	0	0	0	0
0	0	0	0	0	0	1	0	1	0	0	0	0	1	0	1	0	1	0	0	0	0	0	0	0	0	0	0	0	0
3	1	0	0	2	0	4	1	4	1	0	1	3	0	2	3	1	0	1	0	2	4	12	2	0	2	0	0	0	1
0	0	0	0	0	0	0	0	0	0	0	0	0	0	0	0	0	0	0	0	0	0	0	0	0	0	0	0	0	0
1	0	1	1	0	0	1	0	0	0	0	0	2	0	0	0	0	0	0	0	0	0	7	1	2	0	0	0	0	1

C850

0	0	0	0	0	0	0	0	0	0	0	0	0	0	0	0	0	0	0	0	0	0	0	0	0	0	0	0	0	0
0	0	0	0	0	1	0	0	0	1	1	0	0	0	1	0	0	0	1	0	1	0	0	0	0	0	2	1	0	0
0	0	0	0	0	0	0	0	0	0	0	0	0	0	0	0	0	0	0	0	0	0	0	0	0	0	0	0	0	0
0	0	0	0	0	0	0	0	0	0	0	0	0	0	0	0	0	0	0	0	1	0	0	0	0	0	1	0	0	0
0	0	0	0	0	0	0	0	0	0	0	0	0	0	0	0	0	0	0	0	0	0	0	0	0	0	0	0	0	0
0	0	0	0	0	0	0	0	0	0	0	0	0	0	0	0	0	0	0	0	1	0	0	0	0	0	0	0	0	0

C860

0	0	0	0	0	0	0	0	0	0	0	0	0	0	0	0	0	0	0	0	0	0	0	0	0	0	0	0	0	0
0	0	0	0	0	0	0	0	0	0	0	0	0	0	0	0	0	0	0	0	0	0	0	0	0	0	0	0	0	0
0	0	0	0	0	0	0	0	0	0	0	0	0	0	0	0	0	0	0	0	0	0	0	0	0	0	0	0	0	0
0	0	0	0	0	0	0	0	0	0	0	0	0	0	0	0	0	0	0	0	0	1	0	0	0	0	0	0	0	0
0	0	0	0	0	0	0	0	0	0	0	0	0	0	0	0	0	0	0	0	0	0	0	0	0	0	0	0	0	0
0	1	0	0	0	0	0	0	0	0	0	0	0	0	0	0	0	0	0	0	0	0	0	0	0	0	0	0	0	0

C870

0	0	0	0	0	0	0	0	0	0	0	0	0	0	0	0	0	0	0	0	0	0	0	0	0	0	0	0	1	0
1	0	0	0	0	0	0	0	0	0	0	0	0	0	0	0	0	0	0	0	0	0	0	0	0	0	0	0	0	0
0	1	1	0	0	0	0	0	1	0	1	0	0	0	0	0	0	0	0	0	0	0	0	0	0	0	0	0	0	0
0	0	0	0	0	0	0	0	0	0	0	0	1	1	0	0	0	2	0	0	2	1	0	1	0	0	0	0	0	0
0	0	0	0	0	0	0	0	0	0	0	0	0	0	0	0	0	0	0	0	0	0	0	0	0	0	0	0	0	0
0	0	0	0	0	1	0	0	0	0	0	0	0	0	0	0	0	0	1	0	1	0	0	0	0	0	0	0	0	0

C910

1	0	0	0	0	1	1	0	0	0	0	0	0	1	1	0	0	0	0	0	0	0	0	0	0	0	0	0	0	0
0	1	1	1	0	3	0	0	0	1	0	0	0	0	0	0	0	0	0	0	0	0	2	0	0	0	0	0	0	0
0	0	1	0	0	1	0	0	0	0	0	0	0	0	1	0	0	0	0	0	0	0	0	1	0	0	0	0	0	0
1	1	2	0	0	1	0	0	0	0	0	0	1	4	0	1	1	0	0	1	2	0	5	0	0	0	0	0	0	0
0	0	1	2	0	1	0	0	0	0	0	0	0	0	0	0	0	0	0	0	0	0	1	2	0	0	0	0	0	0
1	0	0	1	0	1	0	0	0	0	0	0	2	0	0	0	0	0	0	1	1	0	1	0	0	0	0	0	0	1

C920

0	0	0	0	0	0	0	0	0	0	0	0	0	0	0	0	0	0	0	0	0	0	0	0	0	0	0	0	0	0
0	1	0	1	0	0	0	0	0	0	0	0	0	1	1	0	0	0	0	0	0	1	0	0	1	0	0	0	0	0
0	2	0	0	0	0	0	0	0	1	1	0	0	1	0	0	0	0	0	0	0	0	0	1	0	1	0	0	0	0
0	1	1	0	0	0	0	0	0	0	1	0	0	0	0	1	0	1	0	0	0	0	0	0	0	0	0	0	0	0
1	2	2	3	0	0	1	0	1	1	0	0	0	2	2	0	0	0	0	0	0	0	0	0	0	0	0	0	0	0
0	0	0	0	0	0	0	0	0	0	0	0	0	0	0	0	0	0	0	0	1	0	0	1	0	0	0	0	0	0

C930

0	0	0	0	0	0	1	0	0	0	0	0	1	0	0	0	1	0	0	0	0	0	0	0	0	0	0	1	0	0
1	0	1	0	0	0	0	0	0	0	0	1	0	0	0	0	2	0	0	1	0	2	1	0	2	0	0	0	0	1
0	0	0	0	0	0	0	1	0	1	1	0	1	0	0	0	0	2	0	0	0	0	1	0	0	0	1	0	2	1
0	0	3	0	0	0	0	0	0	0	1	0	3	0	0	3	2	0	0	0	0	0	0	0	0	0	0	1	1	2
0	0	3	0	1	0	0	0	0	0	1	0	0	1	0	0	1	0	0	0	0	0	1	0	0	0	0	0	0	1
0	0	0	0	0	0	0	0	0	0	0	0	2	0	0	0	0	0	0	0	0	0	0	0	0	0	0	2	1	0

C960

0	0	0	0	0	0	0	0	0	0	0	0	0	0	0	0	0	0	0	0	0	0	0	0	0	0	0	0	2	0
0	0	0	0	0	0	0	0	0	0	0	0	0	0	0	0	0	0	0	0	0	0	1	0	1	0	0	0	0	1
1	2	1	0	0	0	1	0	0	0	0	0	1	0	0	1	0	0	1	2	0	0	2	0	0	1	0	0	2	0
0	0	0	0	0	0	0	0	0	0	0	0	5	0	0	0	0	1	3	0	0	0	0	1	1	0	0	0	1	7
1	1	0	0	0	0	1	0	0	0	0	0	0	0	0	0	1	0	0	0	0	0	0	0	0	0	0	1	2	1
0	0	0	0	0	0	0	0	0	0	0	0	0	0	0	0	1	0	0	2	0	0	0	5	0	1	0	0	0	10

C970

0	0	0	0	0	0	0	0	0	0	0	0	0	0	0	0	0	0	0	0	0	0	0	0	0	0	0	0	0	0
1	0	0	0	0	0	0	0	0	0	0	0	0	0	0	0	0	0	0	0	0	0	0	0	1	0	0	0	0	0
3	4	3	3	2	1	7	0	2	0	0	0	1	0	0	0	0	0	0	0	0	0	0	0	0	0	0	0	0	0
0	2	0	0	0	0	0	1	0	0	1	0	0	0	0	0	0	0	0	0	0	0	0	0	0	0	0	0	0	0
0	0	1	0	0	0	0	0	0	0	0	0	0	0	0	0	0	0	0	0	0	0	0	0	0	0	0	0	0	0
0	0	0	0	0	0	0	0	0	0	1	0	0	0	0	0	0	0	0	0	0	0	0	0	0	0	0	0	0	0

C990

0	0	0	0	0	0	0	0	0	0	0	0	0	0	0	0	0	0	0	0	0	0	0	0	0	0	0	0	0	0
0	0	0	0	0	0	0	0	0	0	0	0	0	0	0	0	0	0	0	0	0	0	0	0	0	0	0	0	0	0
0	0	0	0	0	0	0	0	0	0	0	0	1	0	0	0	0	0	0	0	0	0	0	1	0	0	0	0	0	0
0	0	0	0	0	0	0	1	0	0	0	0	0	0	0	0	0	0	0	0	0	0	0	0	0	0	0	0	0	0
0	0	0	0	0	0	0	0	0	0	0	0	0	0	1	0	0	0	1	0	0	0	0	1	0	0	0	0	0	0
0	0	0	0	0	0	0	0	0	0	0	0	0	0	0	0	0	0	0	0	0	0	0	0	0	0	0	0	0	0

APPENDIX 2: General Image System Data for Fifty-One Countries, 1946-1975

C10A

0	0	0	0	0	0	0	0	0	0	0	0	0	0	0	0	0	0	0	0	0	0	0	0	0	0	0	0	0
1	0	0	1	0	0	0	0	0	0	0	0	0	0	0	0	0	0	0	0	0	0	0	0	1	0	0	0	0
0	0	0	0	0	0	0	0	0	0	0	0	0	0	0	0	0	0	0	0	0	0	0	0	0	0	0	0	0
0	0	0	0	0	0	0	0	0	1	0	0	0	0	0	0	0	0	0	0	0	0	0	0	0	0	0	0	0
0	0	0	0	0	0	0	0	0	0	0	0	0	0	0	0	0	0	0	0	0	0	0	0	0	0	0	1	0
0	0	0	0	0	0	0	0	0	0	0	0	0	0	0	0	0	0	0	0	0	0	0	0	1	0	0	0	0

C100A

1	0	0	1	1	0	0	0	0	0	0	0	0	0	0	1	0	1	0	1	0	0	0	0	0	0	0	1	0
0	0	1	1	1	1	0	0	1	0	1	1	0	1	0	0	0	1	0	1	0	0	1	0	1	1	0	1	1
1	0	0	1	1	1	1	1	0	0	0	0	0	0	0	1	0	0	0	0	0	0	0	1	0	1	0	0	0
1	1	0	1	0	1	0	0	0	0	1	1	0	1	1	1	1	0	1	1	0	1	0	0	1	1	0	0	1
1	0	0	1	1	1	0	1	0	0	0	0	1	0	0	0	0	0	1	0	0	0	0	0	1	1	0	1	1
1	0	0	1	0	0	0	0	0	0	1	0	1	0	0	0	0	0	0	0	0	0	0	1	1	0	1	0	1

C1040A

1	1	0	0	0	1	0	0	0	0	0	0	0	0	0	0	0	0	0	0	0	0	0	0	0	0	1	1	0
0	0	0	0	0	0	1	0	0	0	0	0	0	0	1	0	0	0	0	0	0	0	0	1	1	0	0	0	0
1	0	0	1	1	0	1	1	0	0	1	0	1	0	1	0	1	1	0	1	0	0	0	0	0	1	1	0	0
0	0	0	1	1	0	1	0	1	0	0	0	0	0	1	1	0	0	0	0	0	0	0	0	0	0	1	0	1
0	0	0	0	0	0	0	0	0	0	0	0	0	0	0	0	0	0	0	0	0	0	0	0	0	0	0	0	0
1	0	0	0	0	0	1	0	0	0	0	1	1	0	1	0	1	0	0	0	0	0	0	0	1	0	1	0	0

C1060A

1	1	0	0	0	1	0	0	0	0	1	0	1	0	0	0	0	0	0	0	0	1	0	0	1	0	0	0	1
1	0	0	0	1	1	0	0	0	0	1	0	0	0	0	0	0	0	0	0	0	0	0	1	1	0	0	0	1
1	1	0	1	1	0	1	0	1	0	0	1	1	1	1	0	1	0	0	1	0	0	1	1	0	1	0	0	1
0	1	0	0	0	1	0	0	1	0	1	0	0	0	0	0	0	0	0	0	1	1	1	0	1	0	1	0	1
1	0	0	0	0	0	0	0	0	0	0	0	0	0	0	0	0	0	0	0	0	0	0	0	0	0	0	0	0
0	1	0	1	0	0	0	0	0	0	0	0	0	0	0	0	1	0	0	0	0	1	0	1	1	0	0	0	0

C1091A

0	0	0	1	0	0	0	0	0	0	0	0	0	0	0	0	0	0	0	0	0	0	0	0	1	0	0	0	0
0	0	0	0	0	0	0	0	0	0	0	0	0	0	0	0	0	1	0	0	0	0	0	0	0	1	1	0	0
0	0	0	0	0	0	0	0	0	0	0	0	0	0	0	0	0	0	0	0	0	0	0	0	0	0	0	0	0
0	0	0	0	0	0	0	0	0	0	0	0	0	0	0	0	0	0	0	0	0	0	0	0	0	0	0	0	0
0	0	0	0	0	0	0	0	0	0	0	0	0	0	0	0	0	0	0	0	0	0	0	0	0	0	0	0	0
0	0	0	0	0	1	0	1	0	0	0	0	0	0	0	0	0	0	0	0	0	0	0	0	0	0	0	0	0

C1092A

0	0	0	0	0	0	0	0	0	0	0	0	0	0	0	0	0	0	0	0	0	0	0	0	0	1	0	0	0
0	0	0	0	0	0	0	0	0	0	0	1	0	1	0	0	0	0	0	0	0	0	0	0	1	0	0	0	0
0	0	0	0	0	0	0	0	0	0	0	0	0	1	0	0	0	0	0	0	0	0	0	0	0	0	0	0	0
0	0	0	0	0	1	0	0	0	0	0	0	0	0	0	0	0	0	0	0	0	1	1	0	0	0	0	0	0
0	0	0	0	0	0	0	0	0	0	0	0	0	0	0	0	0	0	0	0	0	0	0	0	0	0	0	0	0
0	0	0	0	0	0	0	0	0	0	0	0	0	0	0	0	0	0	0	0	0	0	0	0	0	0	1	0	0

C1100A

0	0	0	0	0	0	0	0	0	0	0	0	0	0	0	0	0	0	0	0	0	0	0	0	0	0	0	0	0
0	0	0	0	0	0	0	0	0	0	0	0	0	0	0	0	0	0	0	0	0	0	0	0	0	0	0	0	0
0	0	0	0	0	0	0	1	0	0	0	0	0	0	0	0	0	0	0	0	0	0	0	0	0	0	0	0	0
0	0	0	0	0	0	0	0	0	0	0	0	0	0	0	0	0	0	0	0	0	0	0	0	0	0	0	0	0
0	0	0	0	0	0	0	0	0	0	0	0	0	0	0	0	0	0	0	0	0	0	0	0	0	0	0	0	0
0	0	0	0	0	0	0	0	0	0	0	0	0	0	0	0	0	0	0	0	0	0	0	0	0	0	0	0	0

*C*1130*A*

0 0 0 0 0 0 0 0 0 0 0 0 0 0 0 0 0 0 0 0 0 0 0 0 0 0 1 0 0 1
1 0 0 1 0 0 0 0 0 0 0 1 0 0 0 0 0 0 0 0 0 0 1 0 1 1 0 1 1 1
1 0 1 1 1 1 0 1 1 0 0 0 1 0 0 1 0 1 0 0 0 0 0 0 0 0 0 1 0 0
0 0 0 1 0 0 0 0 0 0 0 0 0 0 0 0 0 0 0 0 0 0 0 0 0 0 0 1 1 1
0 1 0 1 1 1 1 0 0 0 0 1 1 0 0 0 0 0 0 0 0 0 0 0 0 1 0 1 0 0
0 0 0 0 0 0 0 0 0 0 0 0 0 0 0 0 0 0 0 0 0 0 0 0 1 0 0 1 1 1

*C*1170*A*

0 0 0 0 0 0 0 0 0 0 0 0 0 0 0 0 0 0 0 0 0 0 0 0 0 0 0 0 0 0
0 1 1 1 0 1 0 0 0 0 1 0 1 0 1 0 1 1 0 1 0 0 0 0 1 1 1 1 1 0
1 0 0 0 0 0 1 0 0 1 0 0 1 0 0 1 1 1 0 0 0 0 0 0 0 0 0 1 0 0
1 0 0 0 0 0 0 0 0 1 0 1 0 0 1 0 1 1 0 0 0 1 1 1 1 1 0 0 0 1
0 0 0 0 0 0 0 0 0 0 0 0 0 0 1 0 1 1 0 0 0 0 0 0 0 0 0 0 0 0
0 1 1 0 0 0 0 0 0 0 0 0 0 0 1 0 0 1 0 0 0 0 1 0 1 0 0 0 0 1

*C*1190*A*

0 0 0 0 0 0 0 0 0 0 0 0 0 0 0 0 0 0 0 0 0 0 0 0 0 0 0 0 0 0
0 0 0 0 0 0 0 0 0 0 0 0 0 0 0 0 1 0 0 0 0 0 0 0 1 0 0 0 0 0
0 0 0 1 1 1 1 1 1 1 0 1 0 0 1 0 1 1 1 1 0 0 1 1 0 0 1 1 1 0
0 0 0 0 0 1 0 0 0 0 1 1 0 0 0 1 1 0 0 0 0 0 0 0 0 0 1 0 0 0
0 0 0 0 1 0 0 0 0 0 0 0 0 0 0 0 0 0 0 0 0 0 0 0 0 0 0 0 0 0
0 0 0 0 0 0 0 0 0 0 1 0 0 0 0 0 1 1 0 0 0 0 0 0 1 0 0 0 1 0

*C*120*A*

1 0 0 1 0 0 0 0 1 0 0 0 0 0 0 0 1 1 0 0 0 0 0 0 0 0 0 0 0 0
1 0 1 0 0 1 0 1 1 1 1 0 0 0 0 1 1 1 0 0 0 0 1 1 1 0 0 0 0 0
0 0 0 0 0 0 1 0 0 0 0 0 0 0 0 0 0 0 0 0 0 0 0 1 0 1 0 0 0 0
0 0 1 1 0 1 1 0 1 0 0 0 1 1 1 1 0 0 1 0 0 0 1 0 0 0 0 0 0 0
0 1 0 0 0 1 0 0 1 1 1 0 0 1 0 0 0 1 1 0 0 0 0 0 0 0 0 0 0 0
0 0 0 0 0 0 1 0 1 0 0 0 1 0 0 1 0 0 0 1 1 1 0 1 1 0 0 0 1 0

*C*1210*A*

0 1 0 0 0 0 0 0 0 0 0 0 0 0 0 0 0 0 0 0 0 0 0 0 1 1 1 1 1 0
0 1 0 1 0 1 0 0 0 0 1 0 0 0 0 0 0 1 0 0 0 0 0 0 1 0 0 0 0 0
0 0 0 0 0 0 0 0 0 0 0 0 0 0 0 0 0 0 0 0 0 0 0 0 0 0 0 0 0 0
1 1 0 1 1 1 1 1 1 1 1 1 1 0 0 0 1 1 0 0 0 0 0 1 0 0 0 0 0 0
0 0 0 0 0 0 0 0 0 0 0 0 0 0 0 0 0 0 0 0 0 0 1 0 1 0 0 0 0 0
0 0 0 0 0 0 0 0 0 0 0 0 0 0 1 1 1 1 1 0 0 0 1 0 1 0 0 0 1 0

*C*1220*A*

1 1 0 0 0 0 0 0 0 0 0 0 0 0 0 0 0 0 0 0 0 0 0 0 0 0 0 0 0 0
0 0 0 0 0 0 1 0 0 0 0 0 0 0 0 0 0 0 0 0 0 0 0 0 0 0 0 1 0 0
0 0 1 1 1 1 1 0 1 0 0 0 0 0 0 0 0 0 0 0 0 0 0 0 0 0 0 0 0 0
0 0 0 0 0 0 0 0 0 0 0 0 0 0 1 0 0 1 1 1 1 1 1 0 0 0 0 0 0 0
0 0 0 0 0 0 0 0 0 0 0 0 0 0 0 0 0 0 0 0 0 0 0 0 0 0 0 0 0 0
0 0 0 0 0 0 0 0 0 0 0 0 0 0 1 0 0 1 1 1 1 1 1 1 1 0 0 0 0 0

*C*1240*A*

0 0 0 0 0 1 0 0 0 0 0 0 0 0 1 0 0 0 0 0 0 1 1 0 0 0 1 1 0 0
0 0 1 1 0 0 0 0 0 0 1 0 0 0 0 0 0 0 0 1 0 1 1 1 1 0 1 1 1 0
0 0 0 0 C 0 0 0 0 0 0 1 0 0 0 1 0 0 1 0 0 0 0 1 0 0 0 0 1 0
1 0 0 0 0 0 0 0 0 0 0 0 0 0 0 1 0 0 1 0 0 0 1 1 1 1 0 1 0 0
1 0 0 0 0 0 0 0 0 0 0 0 0 0 0 0 0 0 0 0 0 0 0 0 0 0 0 0 0 0
0 0 0 0 0 0 0 0 0 0 0 0 1 0 1 0 1 0 0 0 0 0 1 0 1 0 0 1 0 0

*C*1250*A*

0	0	0	1	1	0	0	0	0	0	0	0	1	0	0	0	0	0	0	0	0	0	0	0	0	0	0	0	0	0
0	0	1	1	0	0	0	0	0	0	0	0	1	0	1	0	0	1	1	0	0	0	0	0	1	0	0	0	0	0
1	1	1	1	1	0	1	1	1	0	0	1	1	0	1	1	1	0	0	0	0	0	0	0	0	0	0	0	0	0
1	0	0	0	0	1	0	0	0	0	0	0	1	0	1	0	1	0	0	0	1	0	0	1	1	1	0	0	0	0
1	1	1	0	0	1	1	0	0	0	0	0	1	0	1	1	1	0	0	0	1	0	0	0	0	0	0	0	0	0
0	0	0	0	0	0	0	0	0	0	0	1	1	0	1	1	0	0	0	0	0	0	0	1	0	0	0	0	0	0

*C*1290*A*

1	0	0	0	0	0	0	0	0	0	0	0	0	0	0	0	0	0	0	0	0	0	0	0	0	0	0	0	0	0
0	0	0	0	0	0	0	0	0	0	0	0	0	0	0	0	0	0	0	0	0	0	0	0	0	1	0	0	0	0
1	0	1	1	0	1	1	0	0	0	0	0	1	0	0	0	0	0	0	0	0	1	0	0	0	0	1	0	0	1
1	0	0	0	0	0	0	1	0	0	0	0	0	0	0	1	0	0	0	0	0	0	1	0	0	1	0	0	0	0
0	0	0	0	0	0	0	0	0	0	0	0	0	0	0	0	0	0	0	0	0	0	0	0	0	0	0	0	0	0
0	0	0	0	0	0	0	0	1	0	0	0	0	0	0	0	0	0	0	0	0	0	0	1	0	0	0	0	0	0

*C*130*A*

0	0	0	0	0	0	0	0	0	0	0	0	0	0	0	0	0	0	0	0	0	0	0	0	0	0	0	0	0	0
1	0	0	0	0	1	0	0	0	0	0	0	0	0	0	0	0	0	0	0	0	0	0	0	0	0	0	0	0	0
1	1	1	1	1	1	1	0	0	0	1	1	0	0	0	0	1	0	0	1	0	0	0	1	0	0	0	0	0	0
0	0	0	0	0	0	0	0	0	0	0	0	0	0	0	0	0	1	1	0	0	0	0	0	0	0	0	0	0	0
0	1	0	0	0	0	0	0	0	0	0	0	0	0	0	0	0	0	0	0	0	0	0	0	0	0	0	0	0	0
0	0	0	0	0	0	0	0	0	0	0	0	0	0	0	0	0	0	0	0	0	0	0	0	0	0	0	0	0	0

*C*180*A*

0	0	0	0	0	0	0	0	0	0	0	0	0	0	0	0	0	0	0	0	0	0	0	0	1	0	0	0	0	1
0	0	0	0	0	0	0	0	0	0	0	0	1	0	0	0	0	1	0	0	0	0	0	1	1	0	0	0	0	0
0	0	0	0	0	0	0	0	0	0	0	0	0	0	0	0	0	0	0	0	0	0	0	0	0	1	0	0	0	0
1	0	1	1	0	0	0	0	0	1	0	0	0	0	0	0	0	1	1	1	0	0	0	0	1	0	0	0	0	0
0	0	0	0	0	0	0	0	0	0	0	0	0	0	0	0	0	0	0	0	0	0	0	0	0	0	0	0	0	0
0	0	0	0	0	1	0	0	0	0	0	0	1	0	0	0	0	0	0	0	1	0	0	1	1	1	0	0	0	0

*C*20*A*

0	0	0	0	0	0	0	0	0	0	0	0	0	0	0	0	0	0	0	0	0	0	0	0	0	0	0	0	0	0
0	0	0	1	0	1	0	1	0	0	0	0	0	0	0	0	0	0	0	0	0	0	0	0	0	0	0	0	0	0
1	1	1	1	1	1	1	0	0	0	0	0	0	0	0	0	0	0	0	0	0	0	0	0	0	0	0	0	0	0
0	0	0	0	0	0	0	0	0	0	0	0	0	0	0	0	0	0	0	0	0	0	0	0	0	0	0	0	0	0
0	0	0	0	0	1	0	0	0	0	0	0	0	0	0	0	0	0	0	0	0	0	0	0	0	0	0	0	0	0
0	0	0	0	0	0	0	0	0	0	0	0	0	0	0	0	0	0	0	0	0	0	0	0	0	0	0	0	0	0

*C*220*A*

1	1	0	0	1	0	0	0	1	1	1	0	0	0	0	0	0	0	0	1	0	1	0	1	0	0	1	1	0	0
0	1	0	1	0	1	0	0	1	1	0	0	0	0	0	0	0	1	0	0	0	0	1	1	1	1	1	1	0	0
0	1	1	1	0	0	0	0	1	1	1	0	0	0	0	1	0	0	0	0	0	0	0	0	0	1	0	1	1	0
0	0	0	1	0	0	1	0	0	0	0	1	0	0	1	0	0	0	0	1	1	0	0	1	1	0	1	1	0	0
0	0	1	1	0	0	0	0	0	0	0	0	0	0	0	0	0	0	0	0	0	0	0	1	1	0	0	1	0	0
0	0	0	0	0	0	0	0	0	0	0	0	1	0	1	0	0	0	0	1	1	0	0	0	1	1	1	1	1	0

*C*230*A*

0	0	0	0	0	0	0	0	0	0	0	0	0	0	0	0	0	0	0	0	0	1	0	0	0	0	0	0	0	0
1	0	0	0	0	0	0	0	0	0	0	0	0	0	0	0	1	0	0	0	0	1	0	1	1	1	0	0	1	0
0	1	0	0	0	1	1	0	1	1	0	0	1	0	0	0	0	0	0	0	1	0	0	1	0	0	0	0	1	0
1	0	1	0	1	0	0	0	0	0	0	0	0	0	1	0	0	0	0	0	0	1	1	0	0	0	0	0	0	1
1	1	1	1	1	1	1	1	0	0	1	1	0	1	0	0	0	0	0	0	0	0	0	0	0	0	0	0	0	0
0	0	1	0	0	0	0	0	0	0	0	1	0	0	1	0	0	0	0	0	0	1	0	1	0	0	0	0	0	1

*C*240*A*

0	0	0	0	0	0	0	0	0	0	0	0	0	0	0	0	0	0	0	0	0	0	0	0	0	1	0	0	0	0
1	0	0	1	0	0	1	1	0	0	1	1	0	0	0	0	1	1	0	1	0	0	1	0	1	1	0	0	0	0
0	0	0	1	0	0	0	0	1	0	0	1	1	0	0	0	0	1	0	0	0	0	0	0	0	0	0	0	0	0
1	1	1	1	0	0	0	0	1	0	0	1	0	0	0	0	0	0	0	0	0	1	0	1	0	1	0	0	0	0
0	0	1	0	1	0	0	1	0	0	1	1	1	0	0	0	0	0	0	0	0	0	0	0	0	0	0	0	0	0
0	0	0	0	0	0	1	0	0	1	0	0	0	0	0	0	0	1	0	0	0	0	0	0	1	0	0	0	0	1

*C*270*A*

0	1	0	0	0	0	0	0	0	0	0	0	0	0	0	0	0	0	0	0	0	0	0	0	0	0	0	0	0	0
1	0	0	1	0	0	1	0	0	0	0	0	0	0	0	0	0	0	0	0	0	0	0	0	0	0	0	0	0	0
0	0	0	0	1	0	0	0	0	1	0	0	0	0	0	0	0	0	0	0	0	0	0	0	0	0	0	0	0	0
0	1	1	0	1	0	0	0	0	0	0	0	0	0	0	0	0	0	0	0	0	0	0	1	0	0	0	0	0	0
1	0	1	1	0	1	0	0	1	1	0	0	0	0	0	0	0	0	0	0	0	0	0	0	0	0	0	0	0	0
0	1	0	0	0	0	0	0	0	0	0	0	0	0	0	1	0	0	0	0	0	0	0	0	0	0	0	0	0	0

*C*280*A*

0	1	1	0	0	0	0	0	0	0	0	1	0	1	0	0	0	0	0	0	0	0	0	0	0	0	0	0	0	0
0	0	0	1	0	0	1	0	0	0	1	0	0	1	0	0	0	0	0	0	0	0	0	0	1	0	0	0	0	0
0	1	0	1	0	0	1	0	1	0	1	1	1	1	1	1	0	0	0	0	0	0	0	0	0	1	0	0	0	0
1	1	1	1	0	1	0	1	0	1	0	0	1	1	1	1	1	0	0	0	1	0	0	0	0	0	0	0	0	0
1	0	0	0	0	0	1	1	0	0	1	1	1	1	1	1	1	0	0	0	0	0	0	0	0	0	0	0	0	0
0	0	1	0	0	0	1	0	0	0	0	1	1	0	1	1	1	0	0	0	0	0	0	0	0	0	0	0	0	0

*C*320*A*

1	0	0	0	0	0	0	0	0	0	0	0	0	0	0	0	0	0	0	0	0	0	0	0	0	0	0	0	1	0
0	0	0	0	0	0	0	1	0	0	0	1	0	0	0	0	0	0	0	0	1	1	0	0	0	0	0	1	0	0
0	0	0	0	0	0	0	0	0	0	0	0	0	0	0	0	0	0	0	0	0	0	0	0	0	0	0	0	0	0
0	0	0	0	0	1	0	0	0	0	0	0	0	0	0	0	0	0	0	0	0	0	0	0	0	0	0	0	0	0
0	0	0	0	0	0	0	0	0	0	0	0	0	0	0	0	0	0	0	0	0	0	0	0	0	0	0	0	0	0
0	0	0	0	0	0	0	1	0	0	0	0	0	0	0	0	0	0	0	1	0	0	1	0	0	0	1	0	0	0

*C*340*A*

0	0	0	0	0	0	0	0	0	0	0	0	0	1	0	0	0	0	0	0	0	0	0	0	0	0	0	0	0	0
1	0	0	1	0	1	1	0	0	0	0	0	0	1	0	1	1	0	0	1	0	0	0	0	1	0	0	0	0	0
1	0	1	1	0	0	1	0	1	1	0	0	0	0	1	1	1	0	1	0	1	0	0	0	0	0	0	0	0	0
0	0	1	0	0	0	1	1	0	0	0	0	1	1	1	1	1	0	0	1	1	1	0	1	1	0	0	0	0	0
1	1	0	1	0	0	1	0	1	0	0	0	0	0	0	1	1	1	0	0	1	0	0	0	1	1	1	0	0	1
0	0	0	0	0	0	0	0	0	0	0	0	0	0	0	0	0	0	1	1	1	0	0	0	1	0	0	0	0	0

C350A

1 0 0 0 0 0 0 0 0 0 0 0 0 0 0 0 0 0 0 0 0 0 0 0 0 0 0 0 0 0
0 0 0 0 0 1 0 0 0 0 0 0 0 0 0 0 0 0 0 0 0 0 0 0 0 0 0 0 0 0
1 0 0 0 0 0 0 0 0 0 0 0 0 0 1 0 1 0 0 0 0 0 0 0 0 0 0 0 0 0
1 0 0 1 0 0 0 0 0 0 0 0 0 0 0 0 0 0 0 0 1 0 0 0 0 0 0 0 0 0
0 1 1 0 0 0 1 0 0 0 0 0 0 0 1 1 0 0 0 0 0 0 0 0 0 0 1 0 0 0
0 0 0 0 0 0 0 0 0 0 0 0 0 0 0 0 0 0 0 0 1 0 0 0 0 0 1 0 0 0

C380A

0 1 1 0 0 0 0 0 0 0 0 0 0 0 0 0 0 0 0 0 0 0 0 0 0 0 0 0 0 0
1 0 0 1 1 0 1 1 1 0 1 1 1 0 1 1 0 1 1 0 0 0 0 0 1 1 1 0 0 1
1 0 0 1 0 1 0 0 0 0 0 0 0 0 0 0 0 1 0 0 0 0 0 0 0 0 0 0 0 0
0 0 1 1 0 0 0 0 0 0 0 0 0 0 0 0 1 0 0 0 0 0 0 0 0 0 0 0 0 0
0 0 0 0 0 0 0 0 0 0 0 0 0 0 0 0 0 0 0 0 0 0 0 0 0 0 0 0 0 0
0 0 0 0 0 0 0 0 0 0 0 0 0 0 0 0 0 0 0 0 0 0 0 0 0 0 0 0 0 0

C390A

0 1 1 0 1 1 1 1 0 0 0 0 0 0 0 1 0 0 0 0 1 0 0 1 0 0 0 1 1 0
1 1 1 1 1 1 1 1 0 1 0 1 0 0 0 0 1 0 0 0 0 0 0 0 0 0 0 0 0 0
0 0 0 0 1 0 1 1 0 0 0 0 1 1 1 1 1 0 0 0 0 0 1 0 0 0 0 0 0 0
0 1 1 0 0 0 0 0 0 0 1 0 0 0 1 0 0 1 0 0 0 0 1 0 1 0 0 0 1 0
0 1 0 1 0 0 0 0 0 0 0 0 0 0 0 0 0 0 0 0 0 0 0 0 0 0 0 0 0 0
0 1 0 1 1 0 0 0 0 0 0 0 0 0 1 0 1 1 0 0 0 0 1 0 1 0 0 0 1 1

C40A

0 0 0 1 0 1 0 0 0 0 1 1 0 1 0 1 0 0 0 0 0 0 0 1 1 0 0 0 0 1
0 0 0 0 0 0 0 0 0 0 1 0 0 1 1 0 1 1 0 0 0 0 0 0 1 0 0 0 0 1
1 0 0 0 0 1 1 0 1 1 1 1 1 1 0 0 1 1 0 0 0 0 0 1 0 0 0 0 0 0
1 0 0 0 0 0 0 0 1 1 0 0 0 0 0 0 0 0 0 0 0 0 0 1 0 0 0 1 1 1
0 0 0 0 0 0 1 0 0 1 0 0 0 1 1 1 1 1 0 0 1 0 0 0 1 1 0 0 1 1
0 0 0 0 0 0 0 0 1 1 0 1 1 0 0 0 0 0 0 1 1 0 1 0 1 0 1 1 1 1

C450A

1 0 0 0 0 0 0 1 0 0 0 0 0 0 0 0 0 0 0 0 0 0 0 0 0 0 0 0 1 0
0 1 0 1 1 1 1 0 1 0 0 0 1 0 0 1 0 1 0 1 1 1 0 0 0 0 0 1 1 0
1 1 0 0 0 1 1 0 0 0 0 0 0 0 0 0 0 0 0 0 0 1 1 1 1 1 1 1 1 1
1 0 1 0 0 0 1 1 1 0 0 0 0 0 0 0 1 1 0 1 1 1 0 0 0 0 0 0 1 0
0 1 1 1 0 0 0 0 0 0 0 0 0 0 0 0 0 0 0 0 0 1 0 0 0 0 0 1 0 1
0 0 0 0 1 0 1 0 0 0 0 0 0 0 0 0 0 1 0 0 1 0 1 0 0 1 1 1 1 1

C460A

0 0 0 0 1 0 0 0 0 0 0 0 0 0 0 0 0 0 0 0 0 0 0 0 0 0 0 0 0 0
0 0 0 1 0 1 0 0 1 0 1 0 0 0 0 0 1 1 0 0 0 1 0 0 1 0 0 0 1 0
0 0 0 0 0 0 1 0 1 1 1 0 0 0 0 1 0 0 0 0 0 0 0 0 0 0 0 0 0 0
1 1 0 0 1 1 1 1 1 1 1 1 0 0 1 0 0 0 0 1 0 0 0 0 1 0 0 0 1 0
1 0 1 1 0 0 0 1 1 1 0 1 0 0 1 0 1 1 0 0 0 0 0 0 1 0 0 0 0 0
1 0 0 1 0 1 1 0 1 1 1 0 0 0 1 0 0 0 0 0 0 0 0 0 0 1 0 0 1 0

*C*490*A*

0	0	0	0	0	0	0	0	1	0	1	1	0	0	0	0	0	0	0	0	0	0	0	0	0	0	0	0	0	0
1	0	0	1	0	0	0	0	0	0	1	1	0	1	0	0	0	1	0	0	0	0	0	0	0	1	0	0	0	0
0	0	0	1	0	0	0	0	1	0	1	1	1	0	1	0	0	1	1	0	0	0	1	0	1	0	0	0	0	0
1	0	0	0	0	0	0	0	0	0	0	1	0	0	0	1	0	1	0	0	0	0	0	0	1	0	0	0	0	0
1	0	0	0	0	0	0	0	0	0	0	1	1	0	0	0	0	1	0	0	0	0	0	0	1	0	0	0	0	0
1	0	0	1	0	0	0	0	0	0	0	1	0	1	0	1	0	0	0	0	0	0	0	0	0	0	0	0	0	0

*C*50*A*

0	0	0	1	0	1	0	0	0	0	0	0	0	0	0	0	0	0	0	0	0	0	0	0	0	0	0	0	0	0
0	0	0	0	0	0	0	0	0	0	0	0	0	0	0	0	0	0	0	0	0	0	0	1	0	0	0	0	1	1
0	0	0	0	0	0	0	0	0	0	0	0	0	0	0	0	0	0	0	0	0	0	0	0	0	0	0	0	0	0
0	0	1	1	0	0	0	0	0	0	0	1	0	0	1	0	0	0	0	0	0	0	0	1	0	0	0	0	0	0
0	0	0	0	0	0	0	0	0	0	0	0	0	0	0	0	0	0	0	0	0	0	0	0	0	0	0	0	0	0
0	0	0	0	0	0	0	0	0	0	0	0	0	0	0	0	0	0	0	0	0	0	0	1	1	0	0	0	0	0

*C*500*A*

0	0	0	0	0	0	0	0	0	0	0	0	0	0	0	0	0	0	0	0	0	0	0	0	0	0	0	0	0	0
0	0	0	0	0	0	0	1	0	0	0	0	1	0	0	0	0	0	0	0	0	0	0	0	1	0	0	0	0	0
0	0	0	0	0	0	0	1	0	0	0	0	0	1	0	0	0	0	0	0	0	0	0	0	0	0	0	0	0	0
0	0	1	0	0	0	0	1	0	0	1	0	0	0	0	0	0	0	0	0	0	0	0	1	0	0	0	0	0	0
0	1	0	0	0	0	0	0	0	1	1	0	1	0	0	0	1	0	0	0	0	0	0	0	0	0	1	0	0	1
0	0	0	0	0	0	0	0	0	0	0	0	0	0	0	0	0	0	0	0	0	0	0	1	0	0	0	0	0	1

*C*540*A*

1	0	0	0	0	1	0	0	0	0	0	0	0	0	0	0	0	0	0	0	0	0	0	0	0	0	0	0	0	0
1	0	1	1	1	1	1	1	0	0	0	0	0	1	1	1	0	1	0	0	0	0	0	0	0	0	0	0	0	0
1	0	1	1	1	1	1	1	1	0	1	0	1	1	1	1	0	1	0	0	0	0	0	0	0	1	0	0	0	0
1	0	0	0	0	1	1	1	1	0	0	0	0	0	0	1	0	1	0	0	1	0	0	0	0	0	0	0	0	0
1	0	0	1	0	0	1	1	0	0	0	0	0	0	0	0	1	0	0	0	0	0	0	0	0	0	0	0	0	0
0	0	1	1	0	1	1	1	0	0	0	0	0	0	1	1	1	1	0	0	1	0	0	1	0	0	0	0	0	1

*C*570*A*

1	1	1	1	1	0	1	1	1	0	0	0	0	0	0	0	0	0	0	0	0	0	0	1	1	0	1	0	1	0
0	1	0	0	0	0	0	1	1	1	0	1	0	0	1	0	0	1	1	0	0	0	1	0	1	0	1	1	1	0
0	1	0	1	0	1	1	0	0	0	0	0	0	0	0	0	0	0	0	0	0	0	0	0	0	1	0	0	0	0
1	1	1	1	1	0	0	0	0	0	0	0	0	0	1	0	0	0	0	0	0	0	1	0	1	1	1	0	0	1
1	1	0	0	1	0	0	0	0	0	0	0	0	0	0	0	0	0	0	0	0	0	0	0	0	0	0	0	0	0
1	1	0	1	0	1	1	1	0	0	0	0	0	0	0	0	0	1	0	0	0	0	1	0	1	1	1	0	1	1

*C*62*A*

0	0	0	0	0	0	0	0	0	0	1	0	0	0	0	0	0	0	0	0	0	0	0	0	0	0	0	0	0	0
0	1	0	0	0	0	0	1	0	0	1	1	0	0	0	0	0	0	0	0	0	0	0	0	0	0	0	0	0	0
0	1	1	1	1	1	1	1	0	1	1	1	1	0	0	0	1	0	0	1	0	0	0	0	0	0	0	0	0	0
0	1	1	0	0	0	0	0	0	0	1	1	0	0	0	0	0	0	0	0	0	0	0	0	0	0	0	0	0	0
1	1	0	0	0	0	0	0	0	0	1	0	0	0	0	0	0	0	0	0	0	0	0	0	0	0	0	0	0	0
0	0	0	0	1	0	0	0	0	0	1	0	0	0	0	0	0	0	0	0	0	0	0	0	0	0	1	0	0	0

*C*690*A*

0 0 0 0 0 0 0 0 0 0 0 0 0 0 0 0 0 0 0 0 0 0 0 0 0 0 0 0 0 0
0 0 0 0 0 0 0 0 0 0 0 0 0 0 0 0 0 0 0 0 0 0 0 0 0 0 0 0 0 0
0 0 0 0 0 0 0 0 0 0 0 0 0 0 0 1 0 0 0 0 0 0 0 0 0 0 0 0 0 0
0 0 0 0 0 0 0 0 0 0 0 0 0 0 0 1 0 1 0 0 0 0 0 0 0 0 0 0 0 0
0 0 0 0 0 0 0 0 0 0 0 0 0 0 0 0 0 0 0 0 0 0 0 0 0 0 0 0 0 0
0 0 0 0 0 0 0 0 0 0 0 0 0 0 0 0 0 0 0 0 0 0 0 0 0 0 0 0 0 0

*C*720*A*

0 0 0 0 0 0 0 0 0 0 0 0 0 0 0 0 0 0 0 0 0 0 0 0 0 0 0 0 0 0
0 1 0 0 0 0 0 0 0 0 0 0 1 0 0 0 0 0 0 0 0 0 0 0 0 0 0 0 0 0
1 0 0 0 0 0 0 0 0 0 0 0 0 0 0 0 0 0 0 0 0 0 0 0 0 0 0 0 0 0
0 0 0 0 0 0 0 0 0 0 0 0 0 0 0 0 0 0 0 0 0 0 0 0 0 0 0 0 0 0
0 0 0 0 0 0 0 0 0 0 0 0 0 0 0 0 0 0 0 0 0 0 0 0 0 0 0 0 0 0
0 0 0 0 0 0 0 0 0 0 0 0 0 0 0 0 0 0 0 0 0 0 0 0 0 0 0 0 0 0

*C*80*A*

0 1 1 0 1 0 0 0 1 1 0 0 0 0 1 1 0 0 1 0 0 0 0 0 0 0 1 0 0 0
1 0 1 0 1 0 1 0 0 0 0 0 1 0 0 1 0 1 0 1 1 0 1 0 1 0 1 0 1 0
0 0 0 0 0 0 0 0 1 0 0 0 0 0 0 1 0 0 0 0 0 0 0 0 0 0 0 0 0 0
0 1 0 0 1 0 0 1 0 1 0 0 0 0 1 1 1 1 0 1 1 0 1 0 0 0 0 0 0 0
0 0 0 0 0 0 0 0 0 0 0 0 0 0 0 0 0 0 0 0 0 0 0 0 0 0 0 0 0 0
0 0 0 1 0 1 1 0 0 1 0 0 1 0 1 0 0 0 0 0 0 1 0 1 0 0 0 0 0 0

*C*810*A*

1 0 0 0 0 0 1 0 0 0 0 0 1 0 0 0 0 0 0 0 0 0 1 0 0 0 0 0 0 0
0 0 0 0 0 0 0 0 0 0 0 0 0 0 0 0 0 0 0 0 0 0 1 0 0 0 0 0 0 0
0 0 0 0 0 0 1 0 1 0 0 0 0 1 0 1 0 1 0 0 0 0 0 0 0 0 0 0 0 0
1 0 0 0 1 0 1 0 1 0 0 0 1 0 1 1 0 0 0 0 1 1 1 1 0 1 0 0 0 0
0 0 0 0 0 0 0 0 0 0 0 0 0 0 0 0 0 0 0 0 0 0 0 0 0 0 0 0 0 0
1 0 1 1 0 0 1 0 0 0 0 0 1 0 0 0 0 0 0 0 0 0 1 1 1 0 0 0 0 1

*C*850*A*

0 0 0 0 0 0 0 0 0 0 0 0 0 0 0 0 0 0 0 0 0 0 0 0 0 0 0 0 0 0
0 0 0 0 0 1 0 0 0 1 1 0 0 0 1 0 0 0 1 0 1 0 0 0 0 0 1 1 0 0
0 0 0 0 0 0 0 0 0 0 0 0 0 0 0 0 0 0 0 0 0 0 0 0 0 0 0 0 0 0
0 0 0 0 0 0 0 0 0 0 0 0 0 0 0 0 0 0 0 0 1 0 0 0 0 0 1 0 0 0
0 0 0 0 0 0 0 0 0 0 0 0 0 0 0 0 0 0 0 0 0 0 0 0 0 0 0 0 0 0
0 0 0 0 0 0 0 0 0 0 0 0 0 0 0 0 0 0 0 0 1 0 0 0 0 0 0 0 0 0

*C*860*A*

0 0 0 0 0 0 0 0 0 0 0 0 0 0 0 0 0 0 0 0 0 0 0 0 0 0 0 0 0 0
0 0 0 0 0 0 0 0 0 0 0 0 0 0 0 0 0 0 0 0 0 0 0 0 0 0 0 0 0 0
0 0 0 0 0 0 0 0 0 0 0 0 0 0 0 0 0 0 0 0 0 0 0 0 0 0 0 0 0 0
0 0 0 0 0 0 0 0 0 0 0 0 0 0 0 0 0 0 0 0 0 1 0 0 0 0 0 0 0 0
0 0 0 0 0 0 0 0 0 0 0 0 0 0 0 0 0 0 0 0 0 0 0 0 0 0 0 0 0 0
0 1 0 0 0 0 0 0 0 0 0 0 0 0 0 0 0 0 0 0 0 0 0 0 0 0 0 0 0 0

*C*870*A*

0	0	0	0	0	0	0	0	0	0	0	0	0	0	0	0	0	0	0	0	0	0	0	0	0	0	0	0	1	0
1	0	0	0	0	0	0	0	0	0	0	0	0	0	0	0	0	0	0	0	0	0	0	0	0	0	0	0	0	0
0	1	1	0	0	0	0	0	1	0	1	0	0	0	0	0	0	0	0	0	0	0	0	0	0	0	0	0	0	0
0	0	0	0	0	0	0	0	0	0	0	0	1	1	0	0	0	1	0	0	1	1	0	1	0	0	0	0	0	0
0	0	0	0	0	0	0	0	0	0	0	0	0	0	0	0	0	0	0	0	0	0	0	0	0	0	0	0	0	0
0	0	0	0	0	1	0	0	0	0	0	0	0	0	0	0	0	0	1	0	1	0	0	0	0	0	0	0	0	0

*C*910*A*

1	0	0	0	0	1	1	0	0	0	0	0	0	1	1	0	0	0	0	0	0	0	0	0	0	0	0	0	0	0
0	1	1	1	0	1	0	0	0	1	0	0	0	0	0	0	0	0	0	0	0	0	1	0	0	0	0	0	0	0
0	0	1	0	0	1	0	0	0	0	0	0	0	0	1	0	0	0	0	0	0	0	0	1	0	0	0	0	0	0
1	1	1	0	0	1	0	0	0	0	0	0	1	1	0	1	1	0	0	1	1	0	1	0	0	0	0	0	0	0
0	0	1	1	0	1	0	0	0	0	0	0	0	0	0	0	0	0	0	0	0	0	1	1	0	0	0	0	0	0
1	0	0	1	0	1	0	0	0	0	0	0	1	0	0	0	0	0	0	1	1	0	1	0	0	0	0	0	0	1

*C*920*A*

0	0	0	0	0	0	0	0	0	0	0	0	0	0	0	0	0	0	0	0	0	0	0	0	0	0	0	0	0	0
0	1	0	1	0	0	0	0	0	0	0	0	0	1	1	0	0	0	0	0	0	1	0	0	1	0	0	0	0	0
0	1	0	0	0	0	0	0	0	1	1	0	0	1	0	0	0	0	0	0	0	0	0	1	0	1	0	0	0	0
0	1	1	0	0	0	0	0	0	0	1	0	0	0	0	1	0	1	0	0	0	0	0	0	0	0	0	0	0	0
1	1	1	1	0	0	1	0	1	1	0	0	0	1	1	0	0	0	0	0	0	0	0	0	0	0	0	0	0	0
0	0	0	0	0	0	0	0	0	0	0	0	0	0	0	0	0	0	0	0	1	0	0	1	0	0	0	0	0	0

*C*930*A*

0	0	0	0	0	0	1	0	0	0	0	0	1	0	0	0	1	0	0	0	0	0	0	0	0	0	1	0	0	0
1	0	1	0	0	0	0	0	0	0	0	1	0	0	0	0	1	0	0	1	0	1	1	0	1	0	0	0	0	1
0	0	0	0	0	0	0	1	0	1	1	0	1	0	0	0	0	1	0	0	0	0	1	0	0	0	1	0	1	1
0	0	1	0	0	0	0	0	0	0	1	0	1	0	0	1	1	0	0	0	0	0	0	0	0	0	0	1	1	1
0	0	1	0	1	0	0	0	0	0	1	0	0	1	0	0	1	0	0	0	0	0	1	0	0	0	0	0	0	1
0	0	0	0	0	0	0	0	0	0	0	0	1	0	0	0	0	0	0	0	0	0	0	0	0	0	0	1	1	0

*C*960*A*

0	0	0	0	0	0	0	0	0	0	0	0	0	0	0	0	0	0	0	0	0	0	0	0	0	0	0	0	1	0
0	0	0	0	0	0	0	0	0	0	0	0	0	0	0	0	0	0	0	0	0	0	1	0	1	0	0	0	0	1
1	1	1	0	0	0	1	0	0	0	0	0	1	0	0	1	0	0	1	1	0	0	1	0	0	1	0	0	1	0
0	0	0	0	0	0	0	0	0	0	0	0	1	0	0	0	0	1	1	0	0	0	0	1	1	0	0	0	1	1
1	1	0	0	0	0	1	0	0	0	0	0	0	0	0	0	1	0	0	0	0	0	0	0	0	0	0	1	1	1
0	0	0	0	0	0	0	0	0	0	0	0	0	0	0	0	1	0	0	1	0	0	0	1	0	1	0	0	0	1

*C*970*A*

0	0	0	0	0	0	0	0	0	0	0	0	0	0	0	0	0	0	0	0	0	0	0	0	0	0	0	0	0	0
1	0	0	0	0	0	0	0	0	0	0	0	0	0	0	0	0	0	0	0	0	0	0	0	1	0	0	0	0	0
1	1	1	1	1	1	1	0	1	0	0	0	1	0	0	0	0	0	0	0	0	0	0	0	0	0	0	0	0	0
0	1	0	0	0	0	0	1	0	0	1	0	0	0	0	0	0	0	0	0	0	0	0	0	0	0	0	0	0	0
0	0	1	0	0	0	0	0	0	0	0	0	0	0	0	0	0	0	0	0	0	0	0	0	0	0	0	0	0	0
0	0	0	0	0	0	0	0	0	0	1	0	0	0	0	0	0	0	0	0	0	0	0	0	0	0	0	0	0	0

*C*990*A*

0	0	0	0	0	0	0	0	0	0	0	0	0	0	0	0	0	0	0	0	0	0	0	0	0	0	0	0	0	0
0	0	0	0	0	0	0	0	0	0	0	0	0	0	0	0	0	0	0	0	0	0	0	0	0	0	0	0	0	0
0	0	0	0	0	0	0	0	0	0	0	0	1	0	0	0	0	0	0	0	0	0	0	1	0	0	0	0	0	0
0	0	0	0	0	0	0	1	0	0	0	0	0	0	0	0	0	0	0	0	0	0	0	0	0	0	0	0	0	0
0	0	0	0	0	0	0	0	0	0	0	0	0	0	1	0	0	0	1	0	0	0	0	1	0	0	0	0	0	0
0	0	0	0	0	0	0	0	0	0	0	0	0	0	0	0	0	0	0	0	0	0	0	0	0	0	0	0	0	0

APPENDIX 3: List of All Memoryless Samples with Associated Probabilities

0 0 0 0 0 0	.458	1 0 0 0 0 0	.020
0 0 0 0 0 1	.033	1 0 0 0 0 1	.001
0 0 0 0 1 0	.023	1 0 0 0 1 0	.001
0 0 0 0 1 1	.005	1 0 0 0 1 1	.000
0 0 0 1 0 0	.057	1 0 0 1 0 0	.005
0 0 0 1 0 1	.022	1 0 0 1 0 1	.007
0 0 0 1 1 0	.010	1 0 0 1 1 0	.001
0 0 0 1 1 1	.003	1 0 0 1 1 1	.001
0 0 1 0 0 0	.073	1 0 1 0 0 0	.006
0 0 1 0 0 1	.013	1 0 1 0 0 1	.002
0 0 1 0 1 0	.016	1 0 1 0 1 0	.000
0 0 1 0 1 1	.001	1 0 1 0 1 1	.001
0 0 1 1 0 0	.016	1 0 1 1 0 0	.004
0 0 1 1 0 1	.005	1 0 1 1 0 1	.003
0 0 1 1 1 0	.003	1 0 1 1 1 0	.001
0 0 1 1 1 1	.004	1 0 1 1 1 1	.001
0 1 0 0 0 0	.060	1 1 0 0 0 0	.006
0 1 0 0 0 1	.012	1 1 0 0 0 1	.001
0 1 0 0 1 0	.010	1 1 0 0 1 0	.001
0 1 0 0 1 1	.002	1 1 0 0 1 1	.001
0 1 0 1 0 0	.014	1 1 0 1 0 0	.005
0 1 0 1 0 1	.009	1 1 0 1 0 1	.007
0 1 0 1 1 0	.003	1 1 0 1 1 0	.001
0 1 0 1 1 1	.005	1 1 0 1 1 1	.002
0 1 1 0 0 0	.013	1 1 1 0 0 0	.005
0 1 1 0 0 1	.005	1 1 1 0 0 1	.001
0 1 1 0 1 0	.007	1 1 1 0 1 0	.002
0 1 1 0 1 1	.002	1 1 1 0 1 1	.000
0 1 1 1 0 0	.004	1 1 1 1 0 0	.001
0 1 1 1 0 1	.004	1 1 1 1 0 1	.002
0 1 1 1 1 0	.010	1 1 1 1 1 0	.001
0 1 1 1 1 1	.005	1 1 1 1 1 1	.005

APPENDIX 4: State Transition Matrix with Frequencies as Entries

1	0	0	0	0	0	0	0	1	1	0	0	0	0	0	0	1	0	0	0	0	0	0	0	0	0	0	0	0	0	0	0	0	0	0	0	0	0	0	1	0	0	0	0	0	0	0	0	0	0	0	0	0	1	0	0	1	0	0	0	0	0	0	0
0	0	0	0	0	0	0	0	0	0	0	0	0	0	0	1	0	0	0	0	0	0	0	0	0	0	0	0	0	0	0	0	0	0	0	0	0	0	0	0	0	0	0	0	0	0	0	0	0	0	0	0	0	0	0	0	0	0	0	0	0	0	0	0
0	0	0	0	0	0	0	0	0	0	0	0	0	0	0	0	0	0	0	0	0	0	0	0	1	0	0	1	0	0	0	0	0	0	0	0	0	0	0	0	0	0	0	0	0	0	0	0	0	0	0	0	0	0	0	0	0	0	0	0	1	0	0	0
1	0	0	0	0	0	0	0	0	0	0	0	0	0	0	0	0	0	0	0	0	0	0	0	0	0	0	0	0	0	0	0	0	0	0	0	0	1	0	0	0	0	0	0	0	0	0	0	0	0	0	0	0	0	0	0	0	0	0	0	0	0	0	0
0	0	0	0	0	0	0	0	0	0	0	0	0	0	0	0	0	0	0	0	0	0	0	0	0	0	0	0	0	0	0	0	0	0	0	0	0	0	0	0	0	0	0	0	0	0	0	0	0	0	0	0	0	0	0	0	0	0	0	0	0	0	0	0
0	0	0	0	0	0	0	0	0	1	0	0	0	0	0	0	0	0	0	0	0	0	0	0	0	0	0	0	0	0	0	0	0	0	0	0	0	0	0	0	0	0	0	0	0	0	0	0	0	0	0	0	0	0	0	0	0	0	0	0	0	0	0	1
0	0	0	0	0	0	0	0	0	0	0	0	0	0	0	0	0	0	0	1	0	0	0	0	0	0	0	0	0	0	0	0	0	0	0	0	0	0	0	0	0	0	0	0	0	0	0	0	0	0	0	0	0	0	0	0	0	0	0	0	0	0	0	0
2	0	0	0	0	0	0	0	0	0	0	0	0	0	0	1	0	0	0	0	0	0	0	0	0	0	1	0	0	0	0	0	1	0	0	0	0	0	0	0	0	0	0	0	0	0	0	0	1	0	0	0	0	0	0	2	0	0	0	0	0	0	0	0
0	0	0	0	0	0	0	1	0	0	0	0	0	0	0	2	0	0	0	0	0	0	0	0	0	0	0	0	0	0	0	0	0	0	0	0	0	0	0	0	0	0	0	0	0	0	0	0	0	0	0	0	0	0	0	0	0	0	0	0	0	0	0	0
0	0	0	0	1	0	0	0	0	0	0	0	0	0	0	0	1	0	0	0	0	0	0	0	0	0	0	0	0	0	0	0	0	0	0	0	0	0	0	0	0	0	0	0	0	0	0	0	0	0	0	0	0	0	0	0	0	0	0	0	0	0	0	0
0	0	0	0	1	0	0	0	1	0	0	0	0	1	0	0	0	1	0	0	0	1	0	0	0	0	2	0	0	0	0	0	1	0	0	0	0	0	0	0	0	0	0	0	0	0	0	0	2	0	0	0	0	0	0	0	0	0	0	0	0	0	0	0
1	1	0	0	2	0	1	0	0	0	0	0	0	0	0	0	0	1	0	0	0	0	0	0	1	0	0	0	0	0	0	0	0	0	0	0	0	0	0	0	0	0	0	0	0	0	0	0	0	0	0	0	0	0	0	1	0	0	0	0	0	0	0	0
0	0	0	0	1	0	0	0	0	0	0	0	0	0	0	0	0	0	0	0	0	0	0	0	0	0	0	0	0	0	0	0	0	0	0	0	0	0	0	0	0	0	0	0	1	0	0	0	0	0	0	0	0	0	0	0	0	0	0	0	0	0	0	0
0	1	0	0	0	0	0	0	0	0	0	0	0	0	0	0	0	0	0	0	0	0	0	0	0	0	0	0	0	0	0	0	0	0	0	0	0	0	0	0	0	0	0	0	0	0	0	0	1	0	0	0	0	0	0	0	0	0	0	0	0	0	0	0
0	0	0	0	0	0	0	0	0	0	0	0	0	0	0	0	1	0	0	0	0	0	0	0	0	0	0	0	0	0	0	0	0	0	0	0	0	0	0	0	0	1	0	0	0	0	0	0	0	0	0	0	0	0	0	0	0	0	0	0	0	0	0	0
2	0	0	0	2	0	0	0	0	0	0	0	0	0	0	0	0	0	0	0	0	1	0	0	0	0	0	0	0	0	0	0	0	0	0	0	1	0	0	0	0	0	0	0	0	0	0	0	0	1	0	0	0	0	0	0	0	1	0	0	0	0	0	0
0	0	0	0	0	0	0	0	0	0	0	0	0	0	0	0	0	0	0	0	0	0	0	0	0	0	0	0	0	0	0	0	0	0	0	0	0	0	0	0	0	0	0	0	0	0	0	0	0	0	0	0	0	0	0	0	0	0	0	0	0	0	0	0
0	0	1	0	0	0	0	0	0	0	0	0	0	0	0	0	0	0	0	0	0	0	0	0	0	0	0	0	0	0	0	0	0	0	0	0	0	0	0	0	0	0	0	0	0	0	0	0	0	0	0	0	0	0	0	0	0	0	0	0	0	0	0	0
1	0	0	0	0	0	0	0	1	0	0	0	0	0	0	0	1	0	0	0	0	0	0	0	0	0	0	0	0	0	0	0	0	0	0	0	1	0	0	0	0	0	0	0	0	0	0	0	0	0	0	0	0	0	0	0	0	0	1	0	0	0	0	0
0	1	0	0	1	0	1	0	0	0	0	0	0	0	0	0	0	0	0	0	0	0	0	0	0	0	0	0	0	0	0	0	0	0	0	0	0	0	0	0	0	0	0	0	0	0	0	0	0	0	0	0	0	0	0	0	0	0	0	0	0	0	0	0
0	0	0	0	0	0	0	0	0	0	0	0	0	0	0	0	0	0	0	0	0	0	0	0	0	0	1	0	0	0	0	0	0	0	0	0	0	0	0	0	0	0	0	0	0	0	0	0	0	0	0	0	0	0	0	0	0	0	0	0	0	0	0	0
0	0	0	0	0	0	0	0	0	0	0	0	0	0	0	0	0	0	0	0	0	0	0	0	0	0	0	0	0	0	0	0	0	0	0	0	0	0	0	0	0	0	0	0	0	0	0	0	0	0	0	0	0	0	0	0	0	0	0	0	0	0	0	0
0	0	0	0	0	0	0	0	0	1	0	0	0	0	0	0	0	0	0	0	0	0	0	0	0	0	0	0	0	0	0	0	0	0	0	0	0	0	0	0	0	0	0	0	0	0	0	0	0	0	0	0	0	0	0	1	0	0	0	0	0	0	0	0
3	0	0	0	1	0	0	0	1	0	0	0	0	1	0	0	0	0	0	0	0	0	0	0	0	0	0	0	0	0	0	0	0	0	0	0	1	0	0	0	0	0	0	0	0	0	0	0	0	0	0	0	0	0	0	2	0	0	0	0	0	0	0	0
0	0	0	0	0	0	0	0	0	0	0	0	0	0	0	0	0	0	0	0	0	0	0	0	0	0	0	0	0	0	0	0	0	0	0	0	0	0	0	0	0	0	0	0	0	0	0	0	0	0	0	0	0	0	0	0	0	0	0	0	0	0	0	0
0	0	0	0	0	0	0	0	0	0	0	0	0	0	0	0	0	0	0	0	0	0	0	0	0	0	0	0	0	0	0	0	0	0	0	0	0	0	0	0	0	0	0	0	0	1	0	0	0	0	0	0	0	0	0	0	0	0	0	0	0	0	0	0
3	0	0	0	1	0	0	0	0	0	0	0	0	0	0	0	0	0	1	0	0	0	0	0	0	0	0	0	0	0	0	0	1	0	0	0	0	0	0	0	1	0	0	0	1	0	0	0	0	0	0	0	0	0	0	0	0	0	0	0	0	0	0	0
0	0	0	0	1	1	0	0	0	0	0	0	0	0	0	0	1	0	0	1	1	0	0	0	0	0	0	0	0	1	0	0	1	0	0	0	0	0	0	0	0	0	0	0	0	0	0	0	0	0	0	0	0	0	0	0	0	0	0	0	0	0	0	1
0	0	0	0	0	0	0	0	0	0	0	0	0	0	0	0	0	0	0	0	0	0	0	0	0	0	0	0	0	0	0	0	0	0	0	0	0	0	0	0	0	0	0	0	0	0	0	0	0	0	0	0	0	0	0	0	0	0	0	0	0	0	0	0
0	0	0	0	0	0	0	0	0	0	0	0	0	0	0	0	0	0	1	0	0	0	0	0	0	0	0	0	0	0	0	0	0	0	0	0	0	0	0	0	0	0	0	0	0	0	0	0	0	0	0	0	0	0	0	0	0	0	0	0	0	0	0	0
1	0	0	0	0	0	0	0	0	0	0	0	0	0	0	0	0	0	0	0	0	0	0	0	0	0	0	0	0	0	0	0	1	0	0	0	0	0	0	0	0	0	0	0	0	0	0	0	0	0	0	0	0	0	0	0	0	0	0	0	0	0	0	0
10	0	0	0	0	1	0	0	0	0	0	0	1	1	0	0	2	0	1	0	1	1	0	1	1	0	0	0	0	0	1	0	3	0	0	0	0	0	0	0	0	1	0	0	0	1	0	0	0	0	0	1	0	0	0	0	0	0	0	0	0	0	0	1
1	1	0	0	0	0	1	0	0	0	0	0	0	0	0	0	0	0	0	0	0	0	0	0	0	0	0	0	0	0	1	1	1	0	0	0	0	0	0	0	0	0	0	0	0	0	0	0	0	0	0	0	0	0	0	0	0	0	0	0	0	1	0	0
1	0	2	0	0	0	0	0	2	0	1	0	1	1	0	0	1	0	1	0	1	1	0	1	0	0	0	0	0	0	1	0	0	0	0	0	0	0	0	0	0	0	0	0	0	0	0	0	0	0	0	0	0	0	0	0	0	0	1	0	0	0	0	0
1	0	0	0	1	0	0	0	0	0	0	0	0	0	0	0	1	0	0	0	0	0	0	0	1	1	0	0	0	0	0	0	1	0	0	0	0	0	0	0	0	0	0	0	0	0	0	0	0	0	0	0	0	0	0	0	0	0	0	0	0	0	0	0
1	0	0	0	0	0	1	0	0	0	0	0	0	0	0	0	0	0	0	0	0	0	0	0	0	0	0	0	0	0	0	0	0	0	0	1	1	0	0	0	0	0	0	0	0	0	0	0	0	0	0	0	0	1	0	0	0	0	0	0	0	0	0	1
0	0	0	0	1	0	0	0	0	0	1	0	0	0	0	0	0	0	0	0	0	0	0	0	0	1	0	0	0	0	0	0	0	0	0	0	0	0	0	0	0	0	0	0	0	0	0	0	0	0	0	0	0	0	0	0	0	0	0	0	0	0	0	0
1	0	0	0	1	0	0	0	1	0	1	0	1	0	0	0	1	0	0	0	0	0	0	0	0	1	0	0	0	0	0	0	0	0	0	1	0	0	0	0	0	0	0	0	1	0	0	0	0	0	0	0	0	0	0	0	0	0	0	0	0	0	0	0
2	0	0	0	0	0	0	0	0	0	0	0	1	0	0	0	1	0	0	0	1	0	0	1	1	0	0	0	0	0	0	0	0	0	0	0	0	0	0	0	1	0	0	0	0	0	0	0	0	0	0	0	0	0	0	0	0	0	0	0	0	0	0	0
3	0	0	0	3	0	0	0	5	1	0	0	0	0	0	0	1	1	0	0	0	0	0	0	0	0	0	0	1	0	0	0	0	0	0	0	0	0	0	0	0	0	0	0	0	0	0	0	0	0	0	0	0	1	0	0	0	0	0	0	0	0	0	0
2	0	0	0	1	0	0	0	0	1	0	0	0	0	0	0	0	0	0	0	0	0	0	0	0	0	0	0	0	0	0	0	0	0	0	0	0	0	0	0	0	0	0	0	0	0	1	0	0	0	1	0	0	1	0	0	0	0	0	0	0	0	0	0
3	0	0	0	0	0	0	0	0	0	0	0	0	0	0	0	0	0	0	0	0	0	0	0	0	0	0	0	0	0	0	0	0	0	0	0	0	0	0	0	0	0	0	0	0	0	0	0	0	0	0	0	0	0	0	0	0	0	0	0	0	0	0	0
5	0	0	0	3	1	0	0	0	1	0	0	0	0	0	0	0	0	0	0	1	0	0	0	0	0	0	0	1	0	0	1	0	0	0	0	1	0	0	0	0	0	0	0	0	0	0	0	0	0	0	0	0	0	0	0	0	0	0	0	0	0	0	0
6	2	1	1	3	2	0	0	0	0	0	0	0	0	0	0	0	0	0	0	1	2	0	0	0	0	0	0	0	0	0	0	1	0	0	0	1	1	0	0	0	0	0	0	0	0	0	0	0	0	0	0	0	0	0	0	0	0	0	0	0	0	0	0
0	0	1	0	0	0	0	0	0	0	0	0	0	0	0	0	0	0	0	0	0	0	0	0	0	0	0	0	0	0	1	0	0	0	0	0	0	0	0	0	0	0	0	0	0	0	0	0	0	0	0	1	0	0	0	0	0	0	0	0	0	0	0	0
2	1	0	0	0	0	3	0	1	0	0	0	0	0	0	0	0	2	2	0	0	0	0	0	0	1	0	0	0	0	0	0	0	0	0	0	0	0	0	0	0	0	0	0	0	0	0	0	0	0	0	0	0	0	0	1	0	1	0	0	0	0	0	0
8	1	0	0	1	0	0	0	1	0	1	0	1	0	0	0	1	1	1	0	0	0	0	0	1	1	0	0	0	0	0	0	0	0	0	0	0	0	0	0	0	0	0	0	0	0	0	0	0	0	0	0	0	0	0	0	0	0	0	0	0	0	0	0
40	4	1	0	4	1	0	0	3	1	1	0	2	1	3	0	13	3	0	0	2	0	0	0	2	0	0	0	1	0	0	0	3	0	0	0	0	0	0	0	0	0	0	0	0	0	0	0	1	0	0	0	0	1	0	0	0	0	0	0	0	0	0	0
0	0	0	0	0	0	0	0	0	0	0	1	0	1	0	0	0	0	0	0	0	1	0	0	0	0	0	0	0	0	0	1	0	0	0	0	0	0	0	0	0	0	0	1	0	0	0	0	0	0	0	0	0	0	0	0	0	0	0	0	0	0	1	0
1	0	0	0	0	0	0	0	1	0	0	1	0	0	0	0	0	0	0	0	0	0	0	0	0	0	0	0	0	0	0	0	0	0	0	0	0	0	0	0	0	0	0	0	0	0	0	0	0	0	0	0	0	0	0	0	0	0	0	0	0	0	0	0
2	0	0	0	0	1	0	0	1	0	0	0	0	0	0	0	0	0	0	0	0	1	1	0	0	0	0	0	0	0	0	0	0	0	0	0	0	0	0	0	0	0	0	0	0	0	0	0	0	0	0	0	1	0	0	0	0	0	0	0	0	0	0	0
8	0	1	0	2	1	0	0	0	0	2	0	1	0	0	0	0	2	0	0	0	0	0	0	1	0	0	0	0	0	1	1	0	1	0	0	1	0	0	0	0	0	0	0	0	0	0	0	0	0	0	0	1	0	0	0	0	0	0	0	0	0	0	0
0	0	0	0	0	0	0	0	0	0	0	0	0	0	0	0	0	0	0	0	0	0	0	0	0	0	0	0	0	0	0	1	0	0	0	0	0	0	0	0	0	0	0	0	0	0	0	0	0	0	0	0	0	0	0	0	0	0	0	0	0	1	0	0
3	0	2	0	1	0	2	0	2	0	3	0	1	0	1	0	0	0	1	0	0	0	1	1	1	0	0	0	0	0	2	0	0	0	0	0	0	0	0	0	1	0	0	0	0	0	0	0	0	0	0	0	0	0	0	1	0	0	0	0	0	0	0	1
4	1	1	0	1	0	0	0	3	1	0	0	1	0	0	0	1	0	0	0	1	0	0	0	0	0	0	0	1	1	0	0	0	0	0	0	0	0	1	0	0	1	0	0	0	0	0	0	0	0	0	0	1	0	0	0	0	0	0	0	0	0	0	1
33	0	4	1	3	1	1	1	26	4	5	0	3	0	0	0	7	3	1	0	1	0	0	1	2	0	1	0	1	1	2	0	2	0	0	0	0	0	0	1	0	1	0	0	0	0	0	0	0	0	0	0	2	1	1	0	0	0	0	0	0	0	0	0
1	0	0	0	1	0	0	0	0	1	0	0	0	0	1	0	0	0	0	0	0	0	0	0	0	0	0	0	0	0	0	0	0	0	0	0	0	0	0	0	0	0	0	0	0	0	0	0	0	0	0	0	0	0	0	0	0	0	0	0	0	0	0	0
1	0	2	0	1	1	0	0	0	0	0	0	1	0	0	0	0	0	0	0	0	0	0	0	1	1	0	1	1	0	0	0	0	0	0	0	1	0	0	1	0	0	0	0	0	0	0	0	1	0	1	0	0	0	0	0	0	0	0	0	0	0	0	0
12	2	1	0	1	6	0	0	2	0	0	0	0	0	0	0	1	0	0	0	0	0	0	0	2	1	0	0	0	0	0	0	1	0	0	0	0	1	0	0	1	0	0	0	0	0	0	0	1	0	0	0	1	0	0	0	0	0	0	0	0	0	0	0
28	2	2	2	11	2	1	1	10	0	0	0	0	0	0	1	6	1	1	0	7	1	0	0	0	0	0	0	0	1	1	0	1	0	0	0	0	0	0	1	0	0	0	1	0	1	0	0	0	0	0	0	1	0	0	0	0	0	0	1	0	0	0	0
3	1	1	0	0	0	0	1	0	0	0	0	0	0	0	0	0	0	0	1	0	0	0	0	0	0	0	0	0	0	0	0	0	0	0	0	0	0	0	0	0	0	0	0	0	0	0	0	0	0	0	0	0	0	0	0	0	0	0	0	0	0	0	0
13	0	2	0	1	0	0	1	2	0	3	0	1	0	0	0	1	0	0	0	0	0	0	1	1	0	1	0	1	0	1	0	0	0	0	0	0	0	0	0	0	0	0	1	1	0	0	1	0	0	1	0	0	0	0	0	0	0	0	0	0	0	0	0
20	5	1	0	5	2	1	0	2	0	0	0	2	0	0	0	1	0	0	0	2	2	1	0	0	0	0	0	0	1	1	0	1	0	0	0	1	0	0	0	0	0	0	0	0	0	0	1	0	0	0	1	0	0	0	0	0	0	0	0	0	0	0	0
442	22	11	2	32	9	4	1	43	7	7	0	9	0	0	0	46	5	3	1	2	3	1	1	5	2	3	0	2	0	0	2	10	1	0	0	2	2	0	0	1	0	0	0	1	2	0	0	1	0	0	0	3	0	0	0	1	0	0	0	0	0	1	1

References

Anderson, T. and L. Goodman, 1957. 'Statistical Inference About Markov Chains,' *Annals of Mathematical Statistics*, Vol. 28, pp. 89-110.

Armstrong, J. S., 1967. 'Derivation of Theory by Means of Factor Analysis, or Tom Swift and His Electric Factor Analysis Machine,' *American Statistician,* Vol. 17.

Armstrong, J. S. and P. Soelberg, 1968. 'On the Interpretation of Factor Analysis,' *Psychological Bulletin*, Vol. 70.

Ashby, W. R., 1956. *An Introduction to Cybernetics*, Chapman and Hall, London.

Ashby, W. R., 1962. 'Principles of the Self-Organizing System.' In: *Principles of Self-Organization*, H. von Foerster and G. W. Zopf (eds.), Pergamon Press, London.

Ashby, W. R., 1965a. 'Measuring the Internal Informational Exchange in a System,' *Cybernetica*, Vol. VIII, No. 1.

Ashby, W. R., 1965b. 'Constraint Analysis of Many-Dimensional Relations.' In: *Progress in Biocybernetics*, Vol. 2, N. Wiener and J. Schadé (eds.), Elsevier, Amsterdam.

Ashby, W. R., 1967. 'The Set Theory of Mechanism and Homeostasis.' In: *Automaton Theory and Learning Systems*, D. Stewart (ed.), Thompson, Washington.

Ashby, W. R., 1972. 'Systems and Their Informational Measures.' In: Klir, G. J., 1972a.

Atkin, R., 1974a. 'An Approach to Structure in Architectural and Urban Design,' *Environment and Planning*, Vol. 1, pp. 51-67.

Atkin, R., 1974b. *Mathematical Structure in Human Affairs*, Crane-Russak, New York.

Atkin, R., 1977. *Combinatorial Connectivities in Social Systems*, Birkhauser Verlag, Basel.

Azar, E. E. and J. Ben-Dak, 1975 (eds.). *Theory and Practice of Events Research*, Gordon and Breach, New York.

Banks, A. S., 1972. 'Patterns of Domestic Conflict,' *Journal of Conflict Resolution*, March.

Barto, A. and P. Davis, 1977. 'Dyadic Consensus-Competition Models,' School of Advanced Technology, internal communication.

Berrien, F. K., 1968. *General and Social Systems*, Rutgers University Press, New Brunswick.

Billingsley, P., 1961. 'Statistical Methods in Markov Chains,' *Annals of Mathematical Statistics*, Vol. 32, pp. 12-40.

Blalock, H. M., 1961. *Causal Inferences in Nonexperimental Research*, University of North Carolina Press, Chapel Hill.

Blaug, M., 1962. *Economic Theory in Retrospect*, R. D. Erwin, Homewood, Illinois.

Bohr, N., 1949. 'Discussion with Einstein on Epistemological Problems in Atomic Physics.' In: *Albert Einstein: Philosopher-Scientist*, P. Schilpp (ed.), Library of Living Philosophers, Evanston.

Bohr, N., 1958. 'Quantum Physics and Philosophy.' In: Bohr, 1963.

Bohr, N., 1963. *Essays 1958-1962 on Atomic Physics and Knowledge*, John Wiley, New York.

Boudon, R., 1970. 'An Introduction to Lazarsfeld's Philosophical Papers.' In: *Qualitative Analysis*, P. Lazarsfeld (ed.), Allyn and Dacon, Boston, 1972.

Boulding, K. E., 1950. *A Reconstruction of Economics,* John Wiley, New York.

Boulding, K. E., 1955. 'The Malthusian Model as a General System,' *Social and Economic Studies*, Vol. 4, pp. 195-205.

Boulding, K. E., 1965a. *The Image*, University of Michigan Press, Ann Arbor.

Boulding, K. E., 1956b. 'General Systems Theory – the Skeleton of Science,' *Management Science*, Vol. 2, pp. 197-208.

Boyer, C. V., 1968. *A History of Mathematics*, John Wiley, New York.

Bremermann, H. J., 1962. 'Optimization through Evolution and Re-Combination.' In: *Self Organizing Systems*, M. Yavits, G. Jacobi, G. Goldstein (eds.), Spartan Books, Washington.

Bridgeman, P. W., 1959. *The Way Things Are*, Harvard University Press, Cambridge, Massachusetts.

Brinton, C. C., 1952. *The Anatomy of Revolution*, Vintage Books, New York, 1965 (revised edition).

Broekman, J. M., 1971. *Structuralism*, English Translation: D. Reidel, Dordrecht, 1974.

Broekstra, G., 1976. 'Contraint Analysis and Structure Identification,' *Annals of Systems Research*, Vol. 5, pp. 67-80.

Broekstra, G., 1978. 'On the Application of Informational Measures to the Structure Identification Problem With an Example of the Multivariate Analysis of Qualitative Data from a Panel Study.' In: Cavallo, 1978.

Brown, G. S., 1969. *Laws of Form*, George Allen and Unwin, London.

Bruner, J., J. Goodnow, G. Austin, 1956. *A Study of Thinking*, John Wiley, New York.

Buckley, W., 1967. *Sociology and Modern Systems Theory*, Prentice-Hall, Englewood Cliffs, N.J.

Buckley, W. (ed), 1968. *Modern Systems Research for the Behavioral Scientist*, Aldine, Chicago.

Bunge, M., 1969. 'Testability Today.' In: Bunge, 1973.

Bunge, M., 1973. *Method, Model and Matter*, D. Reidel, Dordrecht.

Bunge, M., 1974. 'Things,' *International Journal of General Systems*, Vol. 1, No. 4.

Bunge, M., 1977. 'The GST Challenge to the Classical Philosophies of Science,' *International Journal of General Systems*, Vol. 4, No. 1.

Cartan, H., 1958. 'Nicolas Bourbaki und die Heutige Mathematik,' Arbeitsgemeinschaft fur Forschung des Landes Nordrhein-Westfalen, Heft 76, 1959.

Cassirer, E., 1955. *The Philosophy of Symbolic Form, Vol. 1: Language*, Yale University Press, New Haven.

Castonguay, C., 1972. *Meaning and Existence in Mathematics*, Springer-Verlag, New York.

Cavallo, R. (ed.), 1978. *Recent Developments in Systems Methodology for Social Science Research*, Martinus Nijhoff, Leiden.

Cavallo, R. and M. Conklin, 1977. 'Systems Methodology in Sociology,' *International Journal of Systems Science*, Vol. 8, No. 1.

Cavallo, R. and S. Islam, 1976. 'General Systems Methodology and Large-Scale Systems.' In: *Cybernetics and Modelling and Simulation of Large-Scale Systems,* T. Oren (ed.), Int. Association for Cybernetics, Namur, Belgium.

Cavallo, R. and G. Klir, 1978a. 'A Conceptual Framework for General Systems Problem Solving,' *International Journal of Systems Science*, Vol. 9, No. 2.

Cavallo, R. and G. Klir, 1978b. 'The Structure of Reconstructable Relations: A Comprehensive Study', *Proceedings of the Fourth European Meeting on Cybernetics and Systems Research* (forthcoming).

Cavallo, R. and E. Ziegenhagen, 1978. 'General Systems Modelling of Conflict Within Nations.' In Cavallo, 1978.

Chatfield, C., 1973. 'Statistical Inference Regarding Markov Chain Models,' *Applied Statistics*, Vol. 22, pp. 7-19.

Chatfield, C. and R. Lemon, 1970. 'Analysing Sequences of Behavioural Events,' *Journal of Theoretical Biology*, Vol. 29, pp. 427-445.

Checkland, P. B., 1972. 'Towards a System-Based Methodology for Real-World Problem Solving,' *Journal of Systems Engineering*, Vol. 3.

Checkland, P. B., 1975. 'The Development of Systems Thinking by Systems Practice – A Methodology from an Action Research Program.' In: Trappl and Hanika, 1975.
Churchman, C. W., 1971. *The Design of Inquiring Systems*, Basic Books, New York.
Cioran, E. M., 1964. *La Chute dans Les Temps.* English Translation: *The Fall into Time*, Quadrangle, Chicago, 1970.
Conant, R. C., 1976. 'Laws of Information Which Govern Systems,' *IEEE Transactions on Systems, Man and Cybernetics*, SMC-6, No. 4, pp. 240-255.
Conant, R. C. and W. R. Ashby, 1970. 'Every Good Regulator of a System Must be a Model of that System,' *International Journal of Systems Science*, Vol. 1, No. 2.
Cortés, F., A. Przeworski, J. Sprague, 1974. *Systems Analysis For Social Scientists*, John Wiley, New York.
deBary, W. T., 1960. *Sources of Chinese Tradition*, Columbia University Press, New York.
Deutsch, K. W., 1963. *The Nerves of Government*, The Free Press, New York.
Deutsch, K. W., 1968. *The Analysis of International Relations*, Prentice-Hall, Englewood Cliffs, N.J.
Deutsch, K. W., 1973. 'Quantitative Approaches to Political Analysis: Some Past Trends and Future Prospects.' In: *Mathematical Approaches to Politics*, Alker, Deutsch, Stoetzel (eds.), Jossey-Bass, San Francisco.
Dieudonnè, J., 1964. 'Recent Developments in Mathematics,' *American Mathematical Monthly*, Vol. 71
Digman, J. M., 1966. 'Interaction and Non-linearity in Multivariate Experiments.' In: *Handbook of Multivariate Experimental Psychology*, R. B. Cattell (ed.), Rand McNalley, Chicago.
Dumont, L., 1966. *Homo Hierarchicus*, English Translation: University of Chicago Press, Chicago, 1970.
Durkheim, E., 1893. *The Division of Labor in Society*, Free Press, Glencoe, Illinois, 1942.
Eddington, A., 1934. 'The Decline of Determinism.' Messenger Lecture at Cornell University, reprinted in *New Pathways in Science*, University of Michigan Press, Ann Arbor, 1959.
Edwards, L. P., 1927. *The Natural History of Revolution*, University of Chicago Press, Chicago.
Einhorn, H. J., 1972. 'Alchemy in the Behavioral Sciences,' *Public Opinion Quarterly*, Vol. 36, pp. 367-378.
Eiseley, L., 1958. *Darwin's Century*, Anchor Book Edition, Doubleday, New York, 1961.
Eiseley, L., 1961, *The Man Who Saw Through Time*, Scribners, New York.
Firestone, J. M. and R. W. Chadwick, 1972. 'Social Responsibility and the Study of Social Conflict: Some Thoughts on the Ethical Aspects of Scientific Social Research,' *General Systems Yearbook*, XVII.
Firestone, J. and D. McCormick, 1972. 'An Exploration in Systems Analysis of Domestic Conflict,' *General Systems Yearbook*, XVII.
Florman, S. C., 1976. *The Existential Pleasures of Engineering*, St. Martin's Press, New York.
Forbes, H. D. and E. R. Tufte, 1968. 'A Note of Caution in Causal Modelling,' reprinted in Kirkpatrick, 1974.
Fuchs, W. R., 1967. *Mathematics for the Modern Mind*, Macmillan, New York.
Fuchs, W. R., 1968. *Cybernetics for the Modern Mind*. English Translation: Macmillan, New York, 1971.
Gaines, B. R., 1976a. 'Foundations of Fuzzy Reasoning,' *International Journal of Man-Machine Studies*, Vol. 8, pp. 623-668.
Gaines, B. R., 1976b. 'On the Complexity of Causal Models,' *IEEE Transactions on Systems, Man and Cybernetics*, Vol. 6, No. 1.
Gaines, B. R., 1977. 'System Identification, Approximation, and Complexity,' *International Journal of General Systems,* Vol. 3, Nol. 3.

Gaines, B. R., 1978a. 'Progress in General Systems Research.' In: Klir, 1978a.
Gaines, B. R., 1978b. 'General System Identification: Fundamentals and Results.' In Klir, 1978a.
Gaines, B. R. and L. J. Kohout, 1977. 'The Fuzzy Decade: A Bibliography of Fuzzy Systems and Closely Related Topics,' *International Journal of Man-Machine Studies*, Vol. 9, pp. 1-68.
Gardner, M., 1958. *Logic Machines and Diagrams*, McGraw-Hill, New York.
Georgescu-Roegen, 1966. *Analytical Economics: Issues and Problems*, Harvard University Press, Cambridge, Massachusetts.
Gomez, P., 1978a. 'Organic Problem-Solving in Management.' In: *Progress in/Cybernetics and Systems Research*, R. Trappl (ed.), Hemisphere, Washington.
Gomez, P., 1978b. 'Organic Problem-Solving in Public Administration: A Systems Methodology.' In: *Management Handbook for Public Administration*, J. Sutherland (ed.), Van Nostrand, New York.
Goodman, L., 1962. 'Statistical Models for Analyzing Processes of Change.' *American Journal of Sociology*, Vol. 68, pp. 57-78.
Goodman, L. and W. Kruskal, 1954. 'Measures of Association for Cross-Classifications,' *Journal of American Statistical Association*, Vol. 49, pp. 732-764.
Gordon, R. A. and L. J. Gleser, 1974. 'A Markov Chains Model of Medical Specialty Choice,' *Journal of Mathematical Sociology*, Vol. 3, No. 2, pp. 259-274.
Graham, G., 1971. *Methodological Foundations for Political Analysis,* Xerox College Publishing, Waltham, Mass.
Greenspan, D., 1973. *Discrete Models*, Addison-Wesley, Reading, Mass.
Gurr, T. R. and V. F. Bishop, 1976. 'Violent Nations, and Others,' *Journal of Conflict Resolution*, Vol. XX, No. 1, pp. 79-110.
Hagopian, M. N., 1974. *The Phenomenon of Revolution*, Dodd, Mead, New York.
Hanson, N. R., 1958. *Patterns of Discovery*, University Press, Cambridge.
Hanson, N. R., 1965. 'Notes Toward a Logic of Discovery.' In: *Perspectives on Pierce*, R. J. Bernstein (ed.), Yale University Press, New Haven.
Hayek, F. A., 1955. *The Counter Revolution of Science: Studies on the Abuse of Reason*, Free Press of Glencoe.
Hayek, F. A., 1964. 'The Theory of Complex Phenomena.' In: *The Critical Approach to Science and Philosophy*, M. Bunge (ed.), Free Press of Glencoe, New York.
Hayek, F. A., 1974. *The Pretence of Knowledge* (Nobel Memorial Prize Lecture). In: Hayek, 1975.
Hayek, F. A., 1975. *Full Employment at Any Price?*, The Institute of Economic Affairs, London.
Hazlewood, L., 1976. 'An Appraisal of the Methodology and Statistical Practices used in the Dimensionality of Nations Project.' In: Hoole and Zinnes, 1976.
Heise, D. R., 1975. *Causal Analysis*, John Wiley, New York.
Heisenberg, W., 1958. *Physics and Philosophy*, Harper and Row, New York.
Heisenberg, W., 1959. 'Wolfgang Pauli's Philosophical Outlook.' In: Heisenberg, 1974.
Heisenberg, W., 1974. *Across the Frontiers*, Harper and Row, New York.
Heisenberg, W., 1976. 'The Nature of Elementary Particles,' *Physics Today*, March 1976, pp. 32-39.
Hibbs, D. A., 1973. *Mass Political Violence: A Cross National Causal Analysis*, John Wiley, New York.
Hilton, G., 1976. *Politometrics*, Columbia University Press, New York.
Hoffman, H., 1971. 'Markov Chains in Ethiopia.' In: Kay, 1971.
Holt, R. T. and J. M. Richardson, 1970. 'Competing Paradigms in Comparative Politics.' In: Holt and Turner, 1970.

Holt, R. T. and J. E. Turner (eds.), 1970. *The Methodology of Comparative Research*, The Free Press, New York.
Holton, G., 1973. *Thematic Origins of Scientific Thought*, Harvard University Press, Cambridge, Massachusetts.
Hoole, F. W. and D. A. Zinnes (eds.), 1976. *Quantitative International Politics: An Appraisal*, Praeger, New York.
Huntington, S. P., 1968. *Political Order in Changing Societies*, Yale University Press, New Haven, Connecticut.
Huxley, A., 1954. *The Doors of Perception*, Harper and Row, New York.
James, W., 1868. 'Letter to Ward,' in *The Letters of William James*, Atlantic Monthly Press, Boston, 1920.
James, W., 1907. *Pragmatism: A New Name for Some Old Ways of Thinking*, Longmans Green, New York.
Job, B. L. and C. W. Ostrom, 1976. 'An Appraisal of the Research Design and Philosophy of Science of the Correlates of War Project,' In: Hoole and Zinnes, 1976.
Jones, S. D. and J. D. Singer, 1972. *Beyond Conjecture in International Politics*, F. E. Peacock, Itasca, Illinois.
Josephson, B., 1975. 'Possible Connections Between Psychic Phenomena and Quantum Mechanics,' *New Horizons*, January 1975.
Jung, C. G., 1921. *Psychological Types*, Vol. 6 of The Collected Works of C. G. Jung, Princeton University Press, Princeton, 1971.
Jung, C. G., 1934. *Civilization in Transition*, Vol. 1 of The Collected Works of C. G. Jung, Princeton University Press, Princeton, 1964.
Jung, C. G., 1946. 'The Spirit of Psychology.' In: *Papers from the Eranos Yearbooks, Vol. I: Spirit and Nature*, Pantheon Books, New York, 1954.
Kaplan, A., 1964. *The Conduct of Inquiry: Methodology for Behavioral Science*, Chandler, San Francisco.
Kay, P., 1971. *Explorations in Mathematical Anthropology*, MIT Press, Cambridge, Massachusetts.
Kemeny, J. G., 1961. 'Mathematics Without Numbers.' In: Lerner, 1961b.
Keynes, j. M., 1921. *A Treatise on Probability*, Vol. VIII of The Collected Writings of John Maynard Keynes, St. Martin's, New York, 1973.
Kirkpatrick, S., 1974. *Quantitative Analysis of Political Data*, C. E. Merrill, Columbus.
Klemm, F., 1954. *A History of Western Technology*, English Translation: MIT Press, Cambridge, Massachusetts, 1964.
Klir, G. J., 1969. *An Approach to General Systems Theory*, Von Nostrand, New York.
Klir, G. J., 1970a. 'On the Relation Between Cybernetics and General Systems Theory.' In: *Progress in Cybernetics*, J. Rose (ed.), Gordon and Breach, London.
Klir, G. J., 1970b. 'A Study of Organizations of Self-Organizing Systems,' *Proceedings of 6th International Conference on Cybernetics*, Namur.
Klir, G. J., 1972a (ed.). *Trends in General Systems Theory*, Wiley-Interscience, New York.
Klir, G. J., 1972b. 'The Polyphonic General Systems Theory.' In: Klir, 1972a.
Klir, G. J., 1972c. *Introduction to the Methodology of Switching Circuits*, Van Nostrand, New York.
Klir, G. J., 1975. 'On the Representation of Activity Arrays,' *International Journal of General Systems*, Vol. 2, No. 3, pp. 149-168.
Klir, G. J., 1976. 'Identification of Generative Structure in Empirical Data,' *International Journal of General Systems*, Vol. 3, No. 2, pp. 89-104.
Klir, G. J. (ed.), 1978a. *Applied General Systems Research: Recent Developments and Trends*, Plenum Press, New York.
Klir, G. J., 1978b. 'Computer-Aided Systems Modelling.' In: *System Theory and Ecology*, E. Halfon (ed.), Academic Press, New York.

Klir, G. J., 1978c. 'General Systems Concepts.' In: *Cybernetics: A Source-Book*, R. Trappl (ed.), Hemisphere, Washington, D.C.

Klir, G. J. and M. Marin, 1970. 'A Multimodel and Computer Oriented Methodology for Synthesis of Sequential Discrete Systems,' *IEEE Transactions on Systems Science and Cybernetics*, SSC-6 (1), pp. 40-48.

Klir, G. J. and H. Uyttenhove, 1976. 'Computerized Methodology for Structure Modelling,' *Annals of Systems Research*, Vol. 5.

Klir, G. J. and H. Uyttenhove, 1977. 'On the Problem of Computer-Aided Structure Identification: Some Experimental Observations and Resulting Guidelines,' *International Journal of Man-Machine Systems*.'

Klir, G. J. and M. Valach, 1965. *Cybernetic Modelling*, ILIFFE Books, London.

Krippendorf, K., 1978. 'A Spectral Analysis of Relations.' In Cavallo, 1978.

Kuhn, T. S., 1962. *The Structure of Scientific Revolutions*, University of Chicago Press, Chicago (second edition, 1970).

Lachenmeyer, C. W., 1971. *The Language of Sociology*, Columbia University Press, New York.

LaPalombara, J., 1970. 'Parsimony and Empiricism in Comparative Politics: An Anti-Scholastic View.' In: Holt and Turner, 1970.

Lasswell, H. D., 1961. 'The Qualitative and the Quantitative in Political and Legal Analysis.' In: Lerner, 1961b.

Leiss, W., 1972. *The Domination of Nature*, Beacon Press, Boston.

Leontief, W., 1966. *Essays in Economics: Theories and Theorizing.* Oxford University Press, New York.

Lerner, D., 1961a. 'Introduction, On Quantity and Quality.' In: Lerner, 1961b.

Lerner, D., (ed.), 1961b. *Quantity and Quality: The Hayden Colloquium on Scientific Method and Concept*, Free Press, New York.

Lipsey, R. G., 1963. *An Introduction to Positive Economics*, Weidenfield and Nicolson, London.

Luce, R. D., 1971. 'Similar Systems and Dimensionally Invariant Laws.' *Philosophy of Science*, Vol. 38, No. 2.

Makridakis, S., 1977. 'The Second Law of Systems,' *International Journal of General Systems*, Vol. 4, No. 1.

Mantoux, P., 1928. *The Industrial Revolution in the Eighteenth Century*, Harper Torchbook Edition, Harper and Row, New York, 1962.

Masterman, M., 1970. 'The Nature of a Paradigm.' In: *Criticism and the Growth of Knowledge*, I. Lakatos and A. Musgrave (eds.), Cambridge University Press, Cambridge.

McCulloch, W. S., 1960. 'The Reliability of Biological Systems.' In: *Self-Organizing Systems*, Yavits and Cameron (eds.), Macmillan, New York.

McKelvey, R. and W. Zavoina, 1975. 'A Statistical Model for the Analysis of Ordinal Level Dependent Variables,' *Journal of Mathematical Sociology*, Vol. 4, No. 1, pp. 103-120.

Meehan, E. J., 1965. *The Theory and Method of Political Analysis*, Dorsey Press, Homewood.

Melcher, A. J. (ed.), 1975a. *General Systems and Organization Theory: Methodological Aspects*, Kent State University Press.

Melcher, A. J., 1975b. 'Theory and Application of Systems Theory: Its Promises, Problems, and Realizations.' In: Melcher, 1975a.

Melville, B., 1976. 'Notes on the Civil Applications of Mathematics,' *International Journal of Man-Machine Studies*, Vol. 8, pp. 501-515.

Mesarovic, M. D., D. S. Macko, Y. Takahara, 1970. *Theory of Hierarchical Multilevel Systems*, Academic Press, New York.

Mesarovic, M. D., Y. Takahara, 1975. *General Systems Theory: Mathematical Foundations*, Academic Press, New York.

Miller, G. A., 1967. *The Psychology of Communication*, Penguin Books, Harmondswroth, G.B.
Mowrer, O. H., 1960. *Learning Theory and the Symbolic Processes*, John Wiley, New York.
Nagel, E. and J. R. Newman, 1959. *Godel's Proof*, Routledge and Kegan Paul, London.
Neumann, E., 1954, *The Origins and History of Consciousness*, Princeton University Press, Princeton.
Northrup, F. S. C., 1947. *The Logic of the Sciences and the Humanities*, Macmillan, New York.
Nurmi, H., 1974. 'On the Concept of Complexity and its Relationship to the Methodology of Policy-Oriented Research,' *Social Science Information*, Vol. 13, pp. 55-80.
Owen, P. J., 1975. 'A Mathematical Proof of Gestalt.' In: Trappl and Hanika, 1975, Vol. II.
Papandreou, A., 1958. *Economics as a Science*, J. B. Lippincott, Chicago.
Papini, G., 1920. *Pragmatismo*, Vallecchi, Firenze.
Pareto, V., 1916, *Trattato di Sociologia Generale* (The Mind and Society; A Treatise on General Sociology), Harcout, Brace, New York, 1935.
Pask, G., 1969, 'The Meaning of Cybernetics in the Behavioral Sciences.' In: *Progress in Cybernetics*, J. Rose (ed.), Gordon and Breach, New York.
Pattee, H. H. (ed.), 1973. *Hierarchy Theory: The Challenge of Complex Systems*, George Braziller, New York.
Pattee, H. H., 1976. 'The Role of Instabilities in the Evolution of Control Hierarchies.' In: *Power and Hierarchical Control*, Burns and Buckley (ed.), Sage, London.
Pauli, W., 1946. Letter to C. G. Jung, quoted in Jung, 1946.
Pauli, W., 1952. 'The Influence of Archetypal Ideas on the Scientific Theories of Kepler.' In: *The Interpretation of Nature and the Psyche*, Jung and Pauli, Pantheon Books, New York, 1955.
Peirce, C. S., 1878. 'The Probability of Induction,' reprinted in *Philosophical Writings of Peirce*, Dover, New York, 1955.
Peirce, C. S., c. 1900. 'On the Foundation of Ampliative Reasoning,' (MS660). In: *The Elements of Mathematics*, Vol. III, Carolyn Eisele (ed.), Mouton, The Hague, 1976.
Piaget, J., 1968. *Le Structuralisme* (*Structuralism*), Presses Universitaires de France. English Edition: Harper Torchbooks, New York, 1971.
Piaget, J. 1970a. *Psychology and Epistemology*. English Translation: Viking, New York, 1971.
Piaget, J., 1970b. *Genetic Epistemology*, Columbia University Press, New York.
Piaget, J., 1970c. *Main Trends in Interdisciplinary Research*, Harper Row, New York.
Pichler, F., 1975. *Mathematische Systemtheorie: Dynamische Konstruktionen*, Walter de Gruyter, Berlin.
Pichler, F., 1978. 'On the Decomposition of General Systems: Simulation by Coupling Quotients.' In: Klir, 1978a.
Polanyi, M., 1961. *The Study of Man*, University of Chicago Press, Chicago.
Popper, K., 1974. *Unended Quest.* Revised Edition: Fontana, 1976.
Porter, B., 1976. 'Requisite Variety in the Systems and Control Sciences,' *International Journal of General Systems*, Vol. 2, No. 4, pp. 225-229.
Przeworski, A. and H. Teune, 1970. *The Logic of Comparative Social Inquiry*, John Wiley, New York.
Quine, W. V. O., 1950. 'Identity, Ostension and Hypothesis.' In: Quine, 1963.
Quine, W. V. O., 1963. *From a Logical Point of View.* Original publication: Harvard University Press, 1953. Second Edition: Harper Torchbooks, New York.
Quine, W. V. O., 1969. 'Epistemology Naturalized,' Essay Five in *Ontological Relativity and Other Essays*, Columbia University Press, New York.
Ramsey, F., 1926. 'Truth and Probability.' In: Ramsey, 1931.
Ramsey, F., 1929. 'Philosophy.' In: Ramsey, 1931.

Ramsey, F., 1931. *The Foundations of Mathematics and Other Logical Essays*, Routledge and Kegan Paul, London.
Robbins, L., 1935. *An Essay into the Nature and Significance of Economic Science*, Macmillan, London, second edition revised and extended, 1962.
Rosen, R., 1977. 'Complexity as a System Property,' *International Journal of General Systems*, Vol. 3, No. 4, pp. 227-232.
Rosenblith, W. A., 1961. 'The Quantification of the Electrical Activity of the Nervous System.' In: Lerner, 1961b.
Rotwein, E., 1962. 'Mathematical Economics: The Empirical View and an Appeal for Pluralism,' reprinted in *The Structure of Economic Science*, Prentice-Hall, Englewood Cliffs, 1966.
Rummel, R. J., 1963. 'Dimensions of Conflict Behavior Within and Between Nations,' *General Systems Yearbook*, Vol. 8.
Rummel, R. J., 1965. 'A Field Theory of Social Action With Application to Conflict Within Nations,' *General Systems Yearbook*, Vol. X.
Rummel, R. J., 1967. 'Understanding Factor Analysis,' *Journal of Conflict Resolution*, Vol. 11, pp. 444-480.
Rummel, R. J., 1970. *Applied Factor Analysis*, Northwestern University Press, Evanston.
Rummel, R. J., 1976. 'Comments on the Reviews of the Dimensionality of Nations Project.' In: Hoole and Zinnes, 1976.
Samuelson, P. A., 1963. 'Replies' to 'Professor Samuelson on Theory and Realism,' *American Economic Association Papers and Proceedings,* Vol. 53.
Schumacher, E. F., 1973. *Small is Beautiful: Economics as if People Mattered*, Harper and Row, New York.
Schrodinger, E., 1932. 'Is Science a Fashion of the Times?' In: Schrodinger, 1957.
Schrodinger, E., 1957. *Science, Theory and Man*, Dover, New York.
Shannon, C. E., 1951. 'The Redundancy of English.' In: *Transactions of the Seventh Conference on Cybernetics*, Macy Fndn., New York.
Shannon, C. E. and W. Weaver, 1949. *The Mathematical Theory of Communication*, University of Illinois Press, Urbana.
Simon, H., 1962. 'The Architecture of Complexity,' *Proceedings of the American Philosophical Society*, Vol. 106, pp. 467-482.
Singer, J. D., 1971. *A General Systems Taxonomy for Political Science*, General Learning Press, New York.
Singer, J. D., 1976. 'Rejoinder to the Critiques of the Correlates of War Project.' In: Hoole and Zinnes, 1976.
Sommerhoff, G., 1974. *Logic of the Living Brain*, John Wiley, London.
Stevens, S., 1936. *Psychology: The Propaedutic Science*, Philosophy of Science, Vol. 3, pp. 90-103.
Stewman, S., 1976. 'Markov Models of Occupational Mobility,' *Journal of Mathematical Sociology*, Vol. 4, No. 2, pp. 201-278.
Stigler, G. J., 1955. *The Nature and Role of Originality in Scientific Progress*, reprinted in Stigler, 1965.
Stigler, C. G., 1965. *Essays in the History of Economics*, University of Chicago Press, Chicago.
Tanter, R., 1966. 'Dimensions of Conflict Behavior Within and Between Nations,' *Journal of Conflict Resolution*, Vol. X.
Trappl, R. and F. Hanika (eds.), 1975. *Progress in Cybernetics and Systems Research*, Hemisphere, Washington, D.C.
Tukey, J., 1954. 'Unsolved Problems of Experimental Statistics,' *Journal of the American Statistical Association*, December 1954.
Tustin, A., 1952. 'Feedback,' *Scientific American*, Vol. 187, pp. 48-55.

Uyttenhove, H., 1978. 'Metasystem Identification: A Methodology for Detection in Time-Dependent Systems.' In Klir, 1978a.

Van den Berg, J. H., 1974. *Divided Existence and Complex Society*, Duquesne University Press, Pittsburgh.

Varela, F. J., 1975. 'A Calculus for Self Reference,' *International Journal of General Systems*, Vol. 2, No. 1, pp. 5-24.

Varela, F. J., 1976. 'Not One, Not Two,' *The Co-Evolution Quarterly*, No. 11, pp. 62-67.

Veblen, T., 1919. *The Place of Science in Modern Civilisation and Other Essays*, B. W. Huebsch, New York.

Von Bertalanffy, L., 1968. *General Systems Theory: Foundations, Developments, Applications*, George Braziller, New York.

Von Foerster, H., 1960. 'On Self-Organizing Systems and Their Environments.' In: *Self-Organizing Systems*, M. Yovits and S. Cameron (eds.), Pergamon Press, London.

Von Foerster, H., 1972. 'Responsibilities of Competence,' *Journal of Cybernetics*, Vol. 2, No. 2.

Von Foerster, H., 1973. 'On Constructing a Reality.' In: *Environmental Design Research II*, W. Preiser (ed.), Dowden, Hutchinson and Ross, Stroudsbourg.

Von Foerster, H., 1974. 'Cybernetics of Cybernetics.' In: *Cybernetics of Cybernetics*, B. C. L. Report No. 73.38, Urbana, Illinois.

Von Neumann, J., 1951. 'The General and Logical Theory of Automata.' In: *Cerebral Mechanism in Behavior*, L. A. Jeffress (ed.), John Wiley, New York.

Warfield, J. N., 1973. 'On Arranging Elements of a Hierarchy in Graphic Form,' *IEEE Transactions on Systems, Man and Cybernetics*, Vol. 3, No. 2.

Warfield, J. N., 1976. *Societal Systems: Planning, Policy and Complexity*, John Wiley, New York.

Watanabe, S., 1969. *Knowing and Guessing*, John Wiley, New York.

Weaver, W., 1948. 'Science and Complexity,' *American Scientist*, October 1968.

Weinberg, G. M., 1975. *An Introduction to General Systems Thinking*, John Wiley, New York.

Whorf, B. L., 1940. 'Science and Linguistics.' In: Whorf, 1956.

Whorf, B. L., 1942. 'Language, Mind, and Reality.' In: Whorf, 1956.

Whorf, B. L., 1956. *Language, Thought and Reality*, MIT Press, Cambridge, Massachusetts.

Wilkenfeld, J., 1974. 'Conflict Linkages in the Domestic and Foreign Spheres.' In: Kirkpatrick, 1974.

Winch, P., 1958. *The Idea of a Social Science*, Humanities Press, New York.

Wittgenstein, L., 1953. *Philosophical Investigations*, Basil Blackwell, Oxford.

Wittgenstein, L., 1956. *Remarks on the Foundations of Mathematics*, MIT Press, Cambridge, Massachusetts, 1967.

Wymore, A. W., 1967. *A Mathematical Theory of Systems Engineering: The Elements,* John Wiley, New York.

Wymore, A. W., 1976. *Systems Engineering Methodology for Interdisciplinary Teams*, John Wiley, New York.

Young, A. M., 1976. *The Reflexive Universe*, Delacorte Press.

Zadeh, L. A., 1969. 'The Concepts of System, Aggregate, and State in System Theory.' In: *System Theory*, Zadeh and Polak (eds.), McGraw-Hill, New York.

Zadeh, L. A., 1974. 'A New Approach to System Analysis.' In: *Man and Computer*, Marois (ed.), North-Holland, Amsterdam.

Zeigler, B. P., 1974. 'A Conceptual Basis for Modelling and Simulation,' *International Journal of General Systems*, Vol. 1, No. 4, pp. 213-228.

Zeigler, B. P., 1976. *Theory of Modelling and Simulation*, John Wiley, New York.

Zeigler, B., 1978. 'Structuring the Organization of Partial Models,' *International Journal of General Systems*, Vol. 4, No. 2.

Zeleny, M., 1976. 'Conflict Dissolution,' *General Systems Yearbook*, Vol. XXI.

Subject Index

Name Index